QUARANTINE COSMOLOGY: WHERE EINSTEIN MEETS IMMUTABILITY

THE DPE AXIOM: CORE OPERATING PRINCIPLE

Quarantine Cosmology is an axiomatic framework that interprets the observable universe as a resistive, decaying remainder of a fractured original reality, rather than as the baseline of existence.

Because all truth is from Him, through Him, and to Him, this work operates under a single, non-negotiable axiom: **God is Deliberate, Perfect, and Eternal.**

THE AXIOMATIC INDEPENDENT

The structural validity of Quarantine Cosmology is not contingent upon the consensus of the Hospice Zone. Quarantine Cosmology did not emerge from cultural analysis, literary theory, or comparative mythology, nor did it arise from the bottom-up aggregation of fragmented observations. It emerges from Inherent Ontology. The deductive reasoning that flows from a single, unshakeable certainty that God is Deliberate, Perfect, and Eternal.

A logic system built on a perfect premise is structurally superior to one built on fragmented, time-bound observations. While the closed-loop framework attempts to map reality by measuring the movement of shadows within the Quarantine, the Sovereign Auditor begins with the light source. If the Source is DPE, then any attribute of the current medium that manifests as accidental, decaying, or temporary must be identified as a localized distortion rather than a fundamental truth. A foundational structure defined by succession, decay, or terminality contradicts the meaning of Eternal Perfection.

1. The Rejection of Inductive Fragility

Inductive reasoning, the extrapolation of general conclusions from specific observable instances, functions as a survival tool within a resistive medium. It operates under a constraint where latency, decay, and impedance limit the scope of inference. To construct a cosmology exclusively from induction is to allow the Quarantine to define ontology by its own conditions. Induction remains valid within a closed system for describing behavior signatures, but it is insufficient for determining ontological origin. The Axiomatic Independent therefore begins not with accumulated observation, but with the nature of the Source, whose attributes precede and overrule the behavior of Sin Mass. Consequently, DPE is not inductively derived. It is an axiomatic revelation constraint.

2. The Supremacy of Inherent Ontology

Inherent Ontology is the study of being as it must exist according to the DPE Axiom. It recognizes that the qualities of the Quarantine (distance, decay, and duration) are not properties of existence but symptoms of the Sovereignty Schism. By beginning with the conclusion that God is Deliberate, Perfect, and Eternal, the observer gains the capacity to identify every entropic process as a clinical deviation.

3. The Structural Superiority of the Perfect Premise

A model that accommodates imperfection at its root is a model destined for collapse. If the premise is flawed, the complexity of the math used to support it, the Math Idolatry[1]of the modern age, serves only to obscure the underlying instability. The Axiomatic Independent operates with zero tolerance for systemic contradiction. Any phenomenon that violates the DPE standard is classified as part of the remainder, not part of the Architecture.

4. The Sovereign Auditor as Independent Operator

Because the framework is axiomatically independent, it does not require the permission of the Lineage Machine to function. It provides a systematic recalibration that allows the observer to bypass institutional rebranding and Corrupted Cognition. The Sovereign Auditor does not search for truth within the loop but instead audits the loop using the standard of the Infinum.

---

[1] By Math Idolatry, this work refers not to mathematics itself, but to the elevation of predictive formalism into metaphysical authority.

Quarantine Cosmology

Einstein Meets Immutability

Elvis Sanchez

Imprint: Elvis Sanchez Publishing

AUTHOR'S REFLECTION: The Night the Architecture Unified

I spent years applying the attributes of God to cosmology one at a time. I treated each attribute as a separate lens and asked what the world must be if God is deliberate, or perfect, or eternal. This method revealed patterns that did not match the claims of the guild. A deliberate God would not create empty space. A perfect God would not produce wasted structure. An eternal God would not generate temporary placeholders. These early thought experiments convinced me that the universe was not built on voids, randomness, or symmetry. They showed me that the guild's assumptions were incompatible with the nature of the Creator.

One night I realized that treating the attributes separately was still a concession to human limitation. I was dividing what cannot be divided. The attributes of God are not independent qualities. They are a single reality expressed through human language. That night I unified them into one axiom. God is Deliberate, Perfect, and Eternal. DPE became the generator of the entire architecture.

Once the attributes were unified, the structure of the cosmos became clear. Space became impedance. Time became delay. Matter became coherence concentration. Forces became stress behaviors. Decay became the signature of fracture. The entire system aligned under one indivisible truth. The architecture did not emerge from physics. It emerged from the nature of God, and physics became the field that required correction.

This was the moment Quarantine Cosmology began. It was the moment the world stopped being a puzzle and became a forensic record of a single deviation from a single perfection.

DEDICATION

To Søren Kierkegaard, whose mind pierced the illusions of his age only to be met with mockery and neglect. Labor was conducted in a world incapable of seeing the courage required to confront truth alone. Yet the labor endures, unbroken, waiting for the minds willing to recognize the effort.

The current work is offered in the same spirit to those who will stand on first principles and refuse the inheritance of contradiction and to those willing to see eternity as it is and time for what it remains as an intrusive blemish, a temporary mark upon the face of immutable reality. If the world mocks the framework, the reaction only proves the necessity of the audit. If the framework is ignored, it will only wait patiently, propagating silently, awaiting the readiness of thought unconstrained by consensus, habit, or fear.

To Kierkegaard, the recognition is owed that insight can be lonely, that the path of understanding often meets mockery before embrace, and that truth, once articulated with clarity, remains immortal even if unrecognized. May the current audit find its way into the minds brave enough to confront inherited illusions, and may the structure, like his, stand unyielding long after its first witness has passed.

To the observer an invitation is now extended to enter a world where the foundations are not borrowed, where no premise undergoes inheritance without examination, and where the Light of Eternity serves as the measure against which all else undergoes judgment. Approach with courage. The ground has shifted, and there remains no return to the old illusions.

GLORY TO GOD

Before the mountains were brought forth, or ever thou hadst formed the earth and the world, even from everlasting to everlasting, thou art God. — ***Psalm 90:2***

QUARANTINE COSMOLOGY: EINSTEIN MEETS IMMUTABILITY

The universe is not hiding extraterrestrials from us.
It is hiding us from the universe.— **The Author**

TABLE OF CONTENTS

FOUNDATIONAL AXIOMS OF QUARANTINE COSMOLOGY

The DPE Axiom is the primary diagnostic, but it necessitates a series of secondary axioms that define the behavior of the Quarantine. If God is Deliberate, Perfect, and Eternal, then the following structural laws are absolute:

1. **The Axiom of Non-Neutrality:** Neutrality is a psychological artifact of the fallen mind, not a property of the universe. In a decaying medium, every force has a vector, and every process has a direction. There is no neutral middle ground in a hospice. There is only the rate of collapse and the orientation toward the Source. To claim neutrality is to ignore the directional bias of entropy. Neutrality is the ethics of indecision, used to shield the mind from the offensive clarity of truth.
2. **The Axiom of Asymmetry**[2]: Symmetry, within a fractured medium, functions as a compensatory construct when the Absolute reference point is inaccessible. It is a mathematical aesthetic invented by humans to cope with the absence of the Absolute. Original reality is asymmetrical. One Creator, one creation, one fracture, and one terminal return.

---

[2] See Appendix K for a clinical dissection of the Asymmetry Principle and the reclassification of symmetry as a compensatory stabilization pattern within the Quarantine.

Even within the Quarantine, the physical evidence of fingerprints and snowflakes exposes the Symmetry Artifact. In a truly symmetrical or balanced system, repetition and mirroring would be the rule. Instead, we find that no two fingerprints and no two snowflakes are identical. The radical uniqueness is the residue of the Original Architecture. Perfection does not need a mirror. It is singular. Symmetry only exists as a temporary illusion of balance within the delay of the quarantine.

3. **The Axiom of No New Entities:** A quarantined, decaying medium cannot generate fundamentally new entities. All apparent instances of birth, creation, or emergence are reconfigurations of pre-existing matter, energy, and structure into configurations that accelerate the medium's overall entropic trajectory. Innovation is merely the discovery of new architectures for decomposition.
4. **The Axiom of Temporal Delay (The Fuse)[3]:** Time is not a dimension, a river, or a neutral backdrop. It is the containment protocol, a delay with a countdown, which holds the world in suspension while the final asymmetry approaches. Time is a mercy window that narrows with every passing moment. Time functions as a containment delay and a finite protocol rather than an ontological dimension. The delay exists because the return hasn't happened yet. Time ends because the return will happen.

---

[3] See Appendix C for a forensic analysis of the temporal rupture and the reclassification of time as a localized sequence of decay subordinate to the Infinum.

5. **The Axiom of Sin Mass[4]:** Mass is the clinical measurement of resistance within the medium. Matter is not a fundamental building block but a state of trapped coherence localized by high impedance. Physical laws are the behavior signatures of the resistance, documenting how the medium thwarts original intent.

[4] See Appendix D for a technical audit of Sin Mass as the resistive density of the hospice zone and its function as the primary mechanism of mechanical drag.

The Architecture of Fragmentation — A Commentary

The Source Text: 1 Corinthians 12:12–27 (NIV)

> *Just as a body, though one, has many parts, but all its many parts form one body, so it is with Christ... Now if the foot should say, Because I am not a hand, I do not belong to the body, it would not for that reason stop being part of the body. And if the ear should say, Because I am not an eye, I do not belong to the body, it would not for that reason stop being part of the body. If the whole body were an eye, where would the sense of hearing be? If the whole body were an ear, where would the sense of smell be?... The eye cannot say to the hand, I don't need you! And the head cannot say to the feet, I don't need you!*

---

1. The Text as a Diagnosis of Fragmentation

Paul describes a single body fractured into isolated parts—eyes, hands, ears, feet—each insisting on its own sufficiency. It is a spiritual illustration, but also an exact pathology of today's intellectual landscape.

- Physics behaves like the eye insisting it is the whole body.

- Theology behaves like the hand insisting it is the whole body.
- Philosophy behaves like the ear insisting it is the whole body.
- Cosmology behaves like the head insisting it is the whole body.
- Ethics behaves like the feet insisting they can walk alone.

Each discipline functions as a severed limb, attempting to operate independently of the Source. The result is predictable dysfunction, contradiction, and incoherence. Paul's metaphor is a forensic description of disciplinary amputation.

2. The Text as an Example of Convergence

The irony is structural. The passage that diagnoses fragmentation is itself a living demonstration of unity. Scripture and science, often treated as incompatible domains, reveal themselves as two expressions of the same architecture.

- Scripture describes the body.
- Science describes the parts.
- Quarantine Cosmology describes the system that unifies them.

The text becomes a bridge, a theological articulation of the same structural truth cosmology rediscovers when it stops treating its own limb as the entire organism.

3. The Structural Insight

The passage isn't just about unity. It is unity. It exposes the fracture (disciplinary isolation) while simultaneously modeling the cure (architectural integration). It shows that the Source has always spoken in structural terms and that the disciplines only appear incompatible when they are amputated from one another.

The body metaphor is the blueprint.

PREFACE

The Audit of the Illusion

Most scientific and theological inquiries begin with a fatal error of treating the observable universe as a baseline. They look at the walls of the prison and call them the foundation of the world. They measure the friction of the medium and call it the laws of nature. Such is the hallmark of the Brilliant Fool: the mind that possesses perfect data but an inverted frame.

Quarantine Cosmology does not add another layer of complexity to the existing pile of speculative models. It is a forensic audit that strips away the inherited altitude of consensus and begins with the recognition that the medium we inhabit is not neutral, not original, and not permanent. We are living in a managed, terminal environment (The Hospice Zone) designed for containment and eventual resolution.

The current framework rejects the Symmetry Artifact claim that the universe exists as a stable, self-perpetuating system. The audit further rejects the Neutrality Myth, which imagines that human observation can remain detached from the directional bias of decay. The foundation remains the DPE Axiom: the unshakeable certainty that a Deliberate, Perfect, and Eternal God cannot be the author of a system defined by delay, friction, and randomness.

The paradoxes that haunt modern physics, such as the dual nature of light, the mystery of gravity, and the arrow of time are not mysteries at all. They are structural signatures of a medium under strain and the clinical symptoms of the Fracture.

Once the false premises of the fallen imagination are removed, many long-standing contradictions dissolve. We find no need for Dark Matter or Alien Saviors to account for cosmic anomalies. We need only recognize the Sin Mass that permeates the medium and the countdown that governs the delay.

Those who sense the structural collapse of the modern consensus are already reaching for the original architecture. Quarantine Cosmology functions as a systematic recalibration and restoration of the Eternal Light in a world that has forgotten how to see.

THE JURISDICTION OF THE AUDIT

Quarantine Cosmology does not seek a seat at the table of modern consensus. Secular science, when extended beyond measurement into metaphysical claims, functions as a closed-loop system that treats the localized constraints of the Quarantine as ontological baselines. It mistakes the walls of the hospice for the foundations of the cosmos. To adopt its premises, even in part, is to accept the **Theological Symbiote** compromise. This symbiote is a jurisdictional error in which theological premises are parasitically attached to secular models, reducing the Sovereign to a subordinate battery or a fine-tuner for a machine governed by autonomous laws. This framework refuses that compromise.

Quarantine Cosmology stands alone by establishing three foundational distinctions:

1. **The Rejection of the Closed Loop:** Secular science assumes the universe is a self-contained system governed by its own internal permanence. Quarantine Cosmology recognizes it as a contained system temporary hospice with a terminal countdown. The Brilliant Fool treats the observable universe as the baseline of existence, failing to recognize that what we observe is an isolated medium separated from the perfect Unfallen Universe. The constraints placed upon the medium to prevent the spread of the Fracture are not universal *laws*; they are systemic *flaws*. The difference is a single letter, but it represents an infinity of ontological meaning. These flaws do not function as the foundation of reality. Instead, these constraints are quarantine protocols.

2. **The Inversion of Sovereignty:** Prevailing models operate on a jurisdictional inversion where the Laws of Physics are positioned as the governing constraints to which the nature of God must be reconciled. In Quarantine Cosmology, the physical laws are recognized as secondary procedural rules of the hospice. These laws define the behavior of a medium under strain but do not define the nature of being. The laws are the friction-thresholds of a compromised environment, not the fundamental state of reality.

   These laws function as gauges that do not dictate possibility, but merely measure the specific limits and resistive thresholds of the current quarantine. The presence of Sin Mass introduces a drag that limits velocity, expansion, and duration. What are traditionally observed as universal laws are simply the terminal points where the medium can no longer sustain activity due to internal friction. The laws are a localized byproduct of the quarantine that is unrelated to the Unfallen Universe and functionally unnecessary within its unconstrained architecture..

3. **The End of the God of the Gaps:** The current framework does not look for God in what science does not know. It finds the evidence of the Fracture in exactly what science does know. Where the secular mind measures the speed of light and calls it a universal constant, the Sovereign Auditor sees the speed of the delay. Where the secular mind sees gravity as a fundamental force, we see the impedance of the medium, the weight of Sin Mass.

The present framework does not constitute a debate conducted within a shared framework. The content remains a declaration of a different jurisdiction. By aligning with the DPE Axiom, the inconsistencies of the fallen medium are not resolved. The inconsistencies undergo exposure as necessary diagnostic symptoms of a localized fracture. The audit does not seek data from within the parameters of the rupture. Instead, the process audits the rupture from the jurisdictional perspective of the Unfallen Universe.

METHODOLOGICAL DECLARATION

The conclusions presented within the current audit were not difficult to reach. Speculative daring, mathematical excess, or conceptual novelty remained unnecessary. The findings followed directly from a single axiom grounded in God's immutable attributes. Once immutability, non-contradiction, and eternity functioned as governing constraints rather than as poetic descriptions, the remaining structure assembled itself with minimal resistance.

Simplicity of a foundational nature remains diagnostic rather than an affectation. When first principles are coherent, downstream conclusions narrow rather than proliferate. Complexity, in such cases, serves as evidence of unresolved contradiction carried forward and disguised as sophistication.

The framework proceeds from the DPE Axiom alone. The process does not synthesize existing frameworks, reconcile competing schools, or inherit unexamined premises. The framework refuses, explicitly, to grant metaphysical authority to models whose jurisdiction remains mathematical or instrumental. Utility does not confer truth. Internal consistency does not establish ontology.

Relativistic effects such as clock dilation remain real, measurable, and repeatable. Einstein's mathematics correctly describes these phenomena. The present audit does not dispute the accuracy, power, or brilliance of relativistic calculation when applied to physical measurement. The equations succeed because the math models *relational differences observed under specific physical constraints.*

Einstein earned his place in history through extraordinary insight, but the culture built around his work has elevated interpretation to the level of dogma. Modern physics praises empiricism while defending a category error simply because it is tied to a famous name. That is not science. Science is the pursuit of truth at any cost, even when it requires correcting the giants who came before us. Einstein was right about clocks; he was wrong about time. The measurements stand, the mathematics stands, but the metaphysics built on top of them does not. Respecting Einstein's achievements does not obligate us to preserve his assumptions. It obligates us to examine them.

A ruler that warps due to extreme temperature will produce a different reading than a ruler in a stable environment. The change in the measurement does not indicate that the physical dimensions of the object have altered. The table remains constant. The error occurs when the observer attributes the distortion of the instrument to the nature of the object. Modern cosmology commits this exact error by confusing the behavior of the clock with the nature of time. Clock dilation is a mechanical response to the resistive medium of the hospice zone. It is the warping of the ruler under the thermal stress of sin-mass. To claim that time itself is relative because the clock has slowed is to claim the table has shrunk because the ruler expanded. Clock dilation ≠ time dilation.

The category error occurs at the point where calculation undergoes promotion to ontology. Clock dilation describes how measured intervals differ between reference frames under specific physical constraints, such as motion or gravitational influence. The data does not, by itself, establish that time is relative in its nature, which being itself is frame-dependent, or that reality fractures according to observer state. These conclusions are not derived from the mathematics. The conclusions are appended to the math. The transition from how clocks behave under constraint to what time is remains metaphysical rather than mathematical.

Relativity therefore functions as an extraordinary mathematical achievement applied to physical measurement, not as a description of being itself. Mathematical success demonstrates utility, not ontological authority. Predictive accuracy does not convert a model into metaphysics. Consensus does not sanctify a premise. Longevity does not resolve a contradiction.

The widespread acceptance of relativistic time as a feature of reality rather than as a feature of measurement represents a silent reclassification: a descriptive model was granted explanatory sovereignty. The reclassification was rarely argued for and even less frequently examined. The premise was assumed. Once assumed, the assumption propagated outward into domains where its jurisdiction was never established.

Nowhere were the consequences of the category error more severe than in theology. When relativistic time was absorbed as metaphysical fact, the concept was quietly introduced as a premise for understanding God's relationship to creation. Eternity underwent reframing through temporal relativity. Immutability underwent reinterpretation considering observer-dependence. Discussions regarding God's relation to time proceeded as though time itself possessed ontological primacy. The mentioned moves were not the result of theological necessity, but of an imported assumption whose origin lay in measurement, not being.

The current work rejects the mentioned inheritance entirely. By removing the category error at the level of first principles, the framework avoids piecemeal argument against the theological positions that arose downstream. Those positions remain irrelevant rather than refuted, because the construction rests on a premise not granted here. An alternative platform is established in which God's immutable attributes govern ontology, and time remains subordinate rather than sovereign as Infinum.

Eternity remains the default state of being. Time is subordinate and exists on a different scale entirely. Granting something finite, decaying, and conditioned precedence over eternity constitutes an inversion born of chronological conditioning. Time remains a temporary blemish on the expanse of eternity, a presence whose decay is inevitable. Temporal progression is fleeting and insignificant in the context of an Immutable Being. When the blemish heals, eternity remains whole and unblemished. Time, like all created constraints, leaves no lasting mark on the foundational reality.

The following content is not a debate conducted on shared ground, but a structure developed from a different foundation. The framework stands or falls on internal coherence and fidelity to non-contradiction. Observers seeking reconciliation with relativistic metaphysics will not find it here. Observers willing to examine first principles without inheritance will find that many long-standing paradoxes dissolve rather than demand resolution. The following content is not an argument constructed to persuade. The content is a structural requirement necessitated by the constraints of the DPE Axiom.

INTRODUCTION

The Forensic Framework

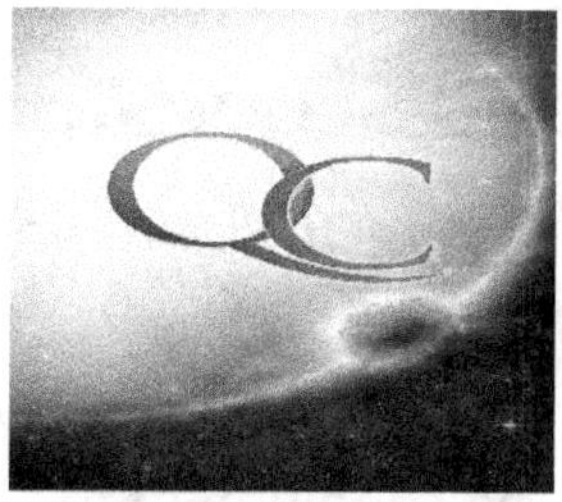

Opening Proclamation

The forensic framework utilized in this work does not constitute a cosmology of origins. The audit serves as a diagnosis of containment and the process proceeds from a single, non-negotiable structural requirement where a Deliberate, Perfect, and Eternal God cannot produce a universe that requires delay, decay, or symmetry to function.

Every subsequent conclusion follows as a clinical consequence of the constraint mentioned. The observable universe does not undergo treatment as a baseline reality. Instead, the cosmos remains the quarantined remainder of a fractured medium. Within the mentioned framework:

- **Light** propagates because coherence collapsed.
- **Time** flows because resistance accumulates.
- **Distance** exists because the medium imposes impedance.
- **Gravity** is the pressure of corruption, not spacetime curvature.

- **Entanglement** is residual conductivity from the unfallen state.
- **The Cosmic Horizon** is a resistance boundary, not an age marker.

Methodology and Scope Constraint

The **Quarantine Cosmology (QC)** framework does not attempt to replace physics or reinterpret Scripture. It operates at a different level of analysis. Physics describes the behavior of the quarantined medium. QC describes the condition of the medium itself.

- **Physics** measures what the medium permits.
- **QC** examines why the medium behaves as it does.

Methodological Rules:

1. **No Assumption of Neutrality:** The observable universe is a damaged environment. Its properties do not reflect original design.
2. **No Default to Ontological Mystery Where Structural Reclassification Is Possible:** Paradoxes (wave-particle duality, entanglement) are interpreted as structural consequences of medium resistance.
3. **No Conflict with Empirical Data:** QC does not dispute measurements but disputes the *interpretations* of those measurements. The data remain the same but the underlying architecture changes.

The Governing Axiom (DPE)

The DPE Axiom defines the boundary conditions of original architecture. If God's nature is taken seriously, the following structural consequences are absolute:

- **Deliberate:** Nothing in the original architecture can be accidental, probabilistic, or dependent on chance. Randomness is a signature of corruption.
- **Perfect:** A perfect Creator would not require inefficiency, friction, or reciprocal tension as foundational necessities of original architecture.
- **Eternal:** An eternal Being cannot create a system that requires time, sequence, or process to exist. Time is a protocol that becomes necessary only when coherence fails.

The Conclusion: If none of these (probability, friction, time) are fundamental, then the observable universe, which depends on all three, cannot be original.

UNIT 1: Ontological Foundations (The Baseline)

CHAPTER 1

The Adhesion of False Premises

1.0 Introduction: Structural Entrapment Within a Corrupted Frame

The mechanics of cognitive entrapment within a quarantined medium begin with a single structural fact. The issue is not error in the conventional sense. It is architectural fixation. Once a false premise is accepted, it becomes the governing interpretive frame. All subsequent data, arguments, and experiences are processed through that frame, regardless of its accuracy. The forensic audit identifies a behavioral signature of fallen cognition where the mind stabilizes around the first coherent frame encountered, even when the construction rests on a corrupted substrate. The adhesion remains structural rather than preferential.

1.1 The Myth of Neutrality: The Non-Existence of a Stable Middle

Neutrality is often treated as an intellectual baseline or a position of methodological equilibrium, yet a structural audit of the medium shows that neutrality requires a static, infinite, and non-resistive environment, none of which characterize the Quarantine. The medium is defined by Directionality. Entropy, temporal duration, decay, and containment pressure operate as irreversible vectors. And in a fractured system every governing force is biased toward collapse. With no point of absolute rest, there is no physical or logical substrate for neutrality. Neutrality is therefore not an architectural property but a psychological artifact secondary effect of the Theological Symbiote that attempts to maintain a conceptual midpoint within a collapsing environment. In forensic terms, neutrality is the interpretive vacuum created when the Absolute (the DPE orientation) is removed from inquiry. Claiming neutrality within a directional hospice is not objectivity but a failure to recognize the terminal bias of the medium.

1.2 Symmetry as a Distortion Artifact

Symmetry arises only when the original reference point has been lost. It is not a marker of perfection but a compensatory construct the mind generates when the true center is inaccessible. In the absence of absolutes, the mind manufactures balance. When perfection cannot be located, it substitutes mirroring. When the Absolute becomes intolerable, it retreats into equilibrium models. Symmetry feels stable because it promises reciprocal forces and mirrored outcomes, yet the medium is not built on reciprocity. Its architecture is directional, irreversible, and non-repeating. A perfect God has no symmetrical opposite, a perfect truth has no neutral midpoint, and a perfect return has no mirrored alternative. Symmetry is not the language of perfection, but a distortion artifact produced by cognition operating without access to the Absolute.

1.3 Asymmetry as the Signature of Singularity

The architecture of reality is asymmetrical. It is defined by a single vector—Creation → Fall → Redemption → Judgment → Eternity—with no cyclical counterpart, mirrored sequence, or equal-and-opposite alternative. The asymmetry is structural: one creation, one incarnation, one cross, one resurrection, one return. A perfect act does not require duplication, a perfect intervention does not require a parallel, and a perfect axis does not require a mirrored counterpart. Asymmetry is not a flaw. It is the signature of singularity within a DPE-anchored system.

1.4 Alien Inversion as a Symmetry Diversion

Certain extraterrestrial narratives function as horizontal reinterpretations of vertical revelation, reclassifying singular intervention as categorical plurality. These narratives attempt to reclassify a singular forensic intervention as one instance within a broader cosmic category. The objective is dilution meant to reduce the absolute nature of the Audit by populating the hospice with hypothetical peers.

1.4.1 The Mechanism of Counterfeit Symmetry

The diversion operates by constructing a false equilibrium between Infinum and the quarantined medium.

- **Symmetrical Balance:** The divine descent is reframed as one example among many possible interstellar visits, reducing the singularity of the Auditor to a category member.
- **Technological Substitution:** Revelation is reframed as advanced technology, implying that the miraculous is merely a higher tier of physics within the Fracture rather than an intervention from outside the medium.

1.4.2 Inversion Disguised as Expansion

Curiosity about extraterrestrial life often conceals a structural dilution of the DPE framework.

- **Vertical to Horizontal Shift:** Vertical revelation (Infinum entering the Quarantine) is reinterpreted as horizontal contact between civilizations.
- **Loss of Jurisdiction:** Once the Auditor is conceptualized as a node among many, His unique jurisdiction over the Fracture is cognitively downgraded.

1.4.3 The Strategic Symmetrization

The alien narrative does not need to deny the Christ-event. It only needs to symmetrize it.

- **Instance vs. Axis:** By placing the Incarnation on a spectrum of cosmic visitors, the absolute axis of the Audit becomes a mere instance.
- **Cloaking Effect:** Creates the illusion of cosmic community within a quarantined hospice, shielding the observer from recognizing the terminal isolation of the medium.

1.5 Premise Adhesion in a Resistive Medium

A premise functions as a load-bearing structural element, and once installed it performs four operations within cognition. First, it filters by determining which data are admitted or excluded. The premise shapes by defining the boundaries of what counts as a plausible explanation. A premise stabilizes by providing equilibrium within a collapsing environment. Finally, it persists by resisting revision because revision is interpreted as structural collapse. In Infinum, the zero-resistance medium, a false premise cannot survive because absolute coherence exposes error immediately. Within the resistive medium, however, Sin Mass introduces delay, allowing contradictions to persist long enough to be normalized, defended, and eventually institutionalized. The mind does not simply hold premises. It adheres to them because the medium itself is adhesive.

1.6 The Self-Healing Error Loop

A false premise generates a closed ecosystem of compensatory errors. Once a corrupted frame is installed, it produces forced interpretations or Bad Outputs that must conform to the frame, and when the frame fails to predict or explain observed behavior, the observer does not abandon it but constructs a Patch. The dynamic produces the **Cycle of Decaytivity**: (1) a premise is adopted, such as the assumption that the universe is original and neutral; (2) contradictions emerge, including missing mass, accelerating expansion, and horizon anomalies; (3) patches are invented—Dark Matter, Dark Energy, Inflation, Alien Interventions; (4) the patches are normalized and reclassified as mysteries or frontier science; (5) the accumulation of patches is treated as evidence of model sophistication; and (6) structural lock-in occurs because questioning the original premise threatens the hierarchy built upon it. The loop is not intellectual failure but the predictable behavior of cognition operating within a resistive medium.

1.7 The Replacement Requirement

A load-bearing frame cannot be removed without a replacement, and argument, ridicule, or data saturation cannot dislodge a premise that stabilizes cognition. Removing a structural beam without providing a superior alternative produces immediate collapse. Therefore, Quarantine Cosmology does not attempt to dismantle the secular frame but replaces it directly: Neutrality → Quarantine, Randomness → Resistance, Emergence → Decaytivity, and Laws → Behavior Signatures. The mind releases a corrupted frame only when a more coherent and more stable frame is available.

1.8 Theological Catastrophe: The Boxed-In God

Theology becomes structurally compromised when it ***fails to distinguish between Observable Science and Speculative Science.*** The difference is the difference between measuring the interior of a containment structure and assuming the containment structure defines the entire ontological domain. Observable Science refers to the clinical recording of behavior signatures within the Hospice Zone. Measurements such as the speed of light, gravitational acceleration, and Clock Dilation describe the symptoms of the quarantine rather than the architecture of Infinum.

Speculative Science emerges when these measurements are reclassified as universal laws, as though the medium's current resistive state were its original design. When theologians adopt such speculative premises, especially those involving Relativity, to appear scientifically aligned, they commit a category error. By treating Time as an ontological dimension rather than a diagnostic of delay, they shrink the Creator into the constraints of the medium.

The result is a transition from a DPE God to a Boxed-In God who must enter time, wait for sequences, and react to temporal processes. This shift reframes the Sovereign Auditor as a participant within the quarantine rather than its Architect and Judge. When the mechanical slowing of a clock is mistaken for the dilation of a dimension, the theologian inadvertently places Infinum under the jurisdiction of the Quarantine.

1.9 Decaytivity: The Compensation Mechanism of Fallen Cognition

Decaytivity is the recombination of decay into explanatory models, emerging when a high-capacity mind attempts to interpret a corrupted environment without acknowledging the fracture. It manifests in the claim that Emergence and Self-Organization are generative forces, when in fact they are turbulence patterns produced by a system exhausting its gradients. Decaytivity is not creativity. It is compensation within a resistive medium.

1.10 Structural Summary

- False premises function as frames, not beliefs.
- Familiarity is mistaken for coherence and consensus is mistaken for certainty.
- Symmetry is constructed when perfection is inaccessible and neutrality is claimed when truth is intolerable.
- Theology collapses when it retrofits God into the behavior signatures of a damaged medium.
- Quarantine Cosmology begins with the indivisible attributes of God—Deliberate, Perfect, Eternal—and treats all observed phenomena as variables of the fracture.

Author's Reflection: My approach to cosmology within the current audit remains top-down. The process does not begin with particles, equations, or inherited models. The starting point remains the attributes of God. The framework proceeds from theological first principles rather than academic consensus. The structure emerged from a refusal to allow a damaged medium to define its Maker. The current volume exists as the architectural consequence of that refusal.

CHAPTER 2

Infinum: The Pre-Fall State (Default Eternity)

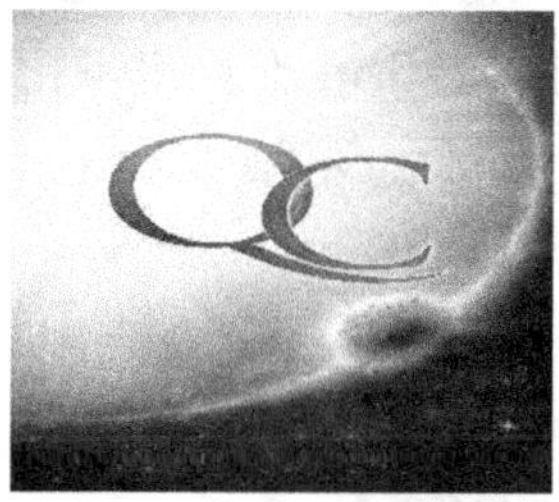

2.0 Introduction

Infinum is the structural condition of reality when the medium contains zero resistance. It is the baseline state of Default Eternity, the immutable substrate that precedes and outlasts the current chronospheric rupture. Infinum is not a coordinate within a relativistic framework. What is traditionally termed spacetime is merely a scarred infrastructure imposed upon Infinum. It is a localized anatomical deviation like a resistive grid necessitated by the Quarantine. The Infinum remains the primary perfection that exists independent of this fractured medium.

2.1 The Properties of Zero-Resistance

Infinum is defined by the total absence of impedance, a condition in which the medium offers no friction, delay, or attenuation to intent or energy. Without the contaminant of Sin Mass, the original architecture operates at full fidelity, allowing identity, energy, and presence to function according to their native design. Where the current observable medium imposes a terminal velocity on light due to the drag of the fracture, the Infinum allows light to simply exist as a state of being. Light in the Unfallen Universe does not traverse, it occupies. It does not travel, it is present.

The properties of Zero Impedance, Absolute Conductivity, and the Absence of Loss and Dissipation describe the operational behavior of a medium in which resistance has not yet been introduced. In this state, the Speed of Light (c) is a nonsensical metric, as there is no resistive medium to calculate a limit against.

2.1.1 Zero Impedance

With no Sin Mass present, the medium imposes no resistance. There is no friction, delay, or attenuation of intent or energy. The medium acts as a perfect conductor for the Deliberate, Perfect, Eternal (DPE) orientation.

2.1.2 Absolute Conductivity

Because the medium lacks resistance, energy does not encounter a barrier. There is no loss of signal strength over distance because distance has not yet been established as a structural cost. It allows for absolute fidelity of information across the entire field.

2.1.3 Absence of Loss and Dissipation

In the zero-resistance state, energy does not dissipate. Coherence is sustained indefinitely without the need for repair, reinforcement, or external energy sources. The system is self-sustaining because it does not fight against a resistive contaminant.

2.2 The Structural Abolition of Metrics

The metrics used to define the physical universe (Distance and Time) are revealed in Infinum as non-operative consequences of impedance.

2.2.1 The Abolition of Distance

Distance is an expression of accumulated resistance. In Infinum, where resistance is zero, distance is non-operative. Separation imposes no cost, meaning that here and there are functionally identical loci of presence.

2.2.2 Zero Traversal

The metrics used to define the physical universe, specifically distance and time, are revealed in Infinum as non-operative consequences of impedance. Measurement is the diagnostic response to a resistive medium. Where there is no resistance, there is no interval; where there is no interval, metrics lack a functional domain.

2.2.3 The Abolition of Time

Time is an accounting protocol for ordering failed coherence. It is a fatal, temporary intrusion, an embolism in the bloodstream of eternity. In a perfectly conductive medium, no ordering is required because there is no latency between intent and manifestation.

2.2.4 The Asymmetry of the Eternal Default

Unlike the Quarantine, which is defined by the arrow of entropy (the terminal countdown of a closed system), Infinum is characterized by the Absolute Asymmetry of Being. Infinum does not possess a past, present, or future to be symmetric about because it is the non-sequential substrate from which all specific manifestations arise.

The introduction of sequence is the introduction of a gap. The introduction of a gap is the introduction of resistance. Therefore, in the zero-resistance medium of the Infinum:

- **Sequence is Non-Operative:** There is no waiting for a result, because there is no distance between the Sovereign intent and the manifestation.
- **Duration is Non-Existent:** Duration is the measurement of friction over a gap. In Infinum, there is no friction and no gap.
- **The Eternal Default:** This is not a "long time." It is the total absence of the temporal metric.

The mistake of the Theological Symbiote is to try to project the improved time of the hospice into the nature of the Sovereign. This framework rejects that. The Sovereign does not exist *through* time. Time is a temporary blemish, like a pimple on Eternity's skin, that exists *upon* the Sovereign's eternity.

2.3 The Formal Logic of the Default (The Formula)

To clinically define Infinum as the baseline, we use the following logical derivation. In a medium where the Sin Mass coefficient (S_{sm}) is zero, the properties of the physical universe dissolve.

$$S_{sm} = 0 \Rightarrow D = 0 \Rightarrow T_r = 0 \Rightarrow T = 0$$

- $\boldsymbol{S_{sm}}$: Sin Mass (Resistance/Impedance).
- $\boldsymbol{D}$: Distance (Integrated Impedance).

- T_r: Traversal (The requirement of movement).
- T: Time (The accounting protocol for delay).

2.3.1 Interpretation of the Formula

The formula states that if Resistance is zero, Distance becomes non-operative. If Distance is non-operative, Traversal (the cost of moving from point A to B) becomes zero. If Traversal is zero, Time (the bookkeeping system for that delay) is rendered obsolete.

2.4 Behavior Signatures of the Default Architecture

The physics of the Infinum represent the baseline operational standards of a medium that has never been subjected to the anatomical deviation of the fracture. Because this architecture is defined by Absolute Asymmetry and Zero Impedance, the resulting behavior signatures do not rely on the compensatory mechanisms found within the hospice zone. In the Quarantine, energy must react to the pressure of the Sin Mass, while in the Infinum, energy simply expresses the intent of the Sovereign without the necessity of survival-based adaptations.

2.4.1 Presence without Propagation

Light does not propagate in Infinum. Propagation is a response to resistance. It is the speed at which a signal fights through a medium. In the Default Eternity, light is simply present. It does not travel from a source. It is the source manifest across the entire field simultaneously.

2.4.2 Energy without Localization

Localization is a collapse event caused by impedance. In the fractured medium, energy is forced to pelletize into particles to survive the pressure of the Sin Mass. In Infinum, energy and identity are global, unconstrained, and non-localized.

2.4.3 Dissolution of Duality

Wave-particle duality is not a fundamental property of matter. It is a structural fault of a corrupted medium. It is the signature of energy trying to exist in a high-pressure environment. In Infinum, energy exists in a singular, stable state of Being.

2.5 The Logic of the Baseline

The Fall did not introduce new laws. It introduced Resistance. The resistance forced the original architecture to behave as if it were governed by delay, decay, and distance.

2.5.1 The Sabbath as a Forensic Portal

The Sabbath is a structural window into Infinum. It is a 24-hour period where the mechanisms of decay (deadlines, production, and schedules) are technically abolished. It allows the identity topology to recalibrate its orientation back toward the Eternal Now.

2.5.2 The Resurrection Witness

The resurrection of the Source-Manifest (Jesus) was the revelation of Infinum breaking through the timeline. It proved that the original, zero-resistance architecture remains the dominant reality, temporarily suppressed but not destroyed by the Quarantine.

2.6 Summary

Infinum is the identity of eternity itself, the primordial substrate that precedes the fracture.

- **Logic**: Qualitative (Presence)
- **Energy**: Perfect (Sustained/Deliberate)
- **Space**: Zero-Latency (Global Presence)
- **Medium**: Zero Impedance ($S_{sm} = 0$)

CHAPTER 3

No New Entities: Cosmology in a Hospice Zone

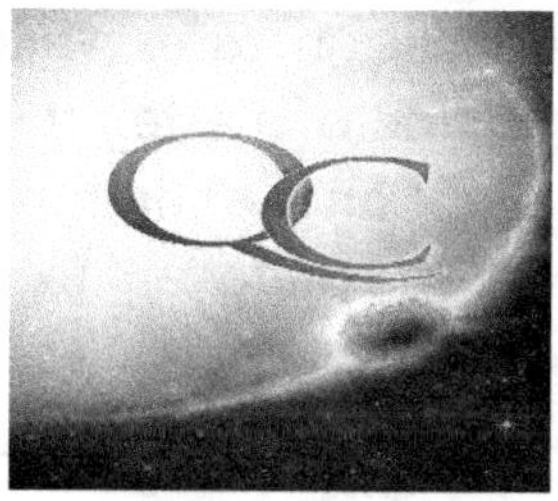

3.0 Introduction

In Quarantine Cosmology, the observable universe is treated not as a fertile or self-renewing cosmos, but as a Hospice Zone: a sealed, terminal environment in which all processes are governed by decay, collapse, and the exhaustion of finite gradients. Within such a medium, the appearance of novelty is an illusion generated by rearrangement, not creation.

The central axiom governing all dynamics inside the Quarantine is the **No New Entities Principle**: A quarantined, decaying medium cannot generate fundamentally new entities. All apparent instances of birth, creation, or emergence are reconfigurations of re-existing matter, energy, and structure into configurations that accelerate the medium's overall entropic trajectory.

The Forensic Definition of an Entity

An entity is a primary Identity Class established within Default Eternity and authorized during the Seven-Day Template. In a forensic context, the classification of an entity relies upon the identification of its jurisdictional origin rather than its current appearance. This diagnostic framework ensures that emergent phenomena are correctly categorized as either primary signatures or derivative distortions.

- **Substance over Simulation:** An entity possesses an original jurisdictional signature. While the Fracture allows for the reconfiguration of matter, these are derivative states. A reconfiguration is a forensic mask, not a new ontological arrival. The material flux of the Quarantine cannot generate a new primary class of being.
- **Topology vs. Ontology:** Emergent properties are topological shifts. Just as a deck of cards produces different hands without adding new cards, the medium produces novel configurations without adding new entities. The arrangement is new but the essence is not. Diversity in the hospice zone is a rearrangement of existing sin mass or existing spirit, not an expansion of the creation.
- **Identity Class:** An entity is a singular category of being authorized during the Seven-Day Template. Any phenomenon appearing after the close of the creation window, whether biological, digital, or perceived as extraterrestrial, is a parasitic variation or a mask of an extant identity. The forensic limit of reality was set at the completion of the seventh day and the Quarantine can only shuffle or corrupt, it cannot originate.

3.1 The Hospice Zone as a Decaying Medium

The Hospice Zone is a sealed environment defined by exhaustion rather than expansion. Its medium no longer generates coherence.

It redistributes the remnants of an earlier order under increasing resistance. Every process within it (physical, biological, technological, or civilizational) operates under the same constraint. The system is running on inherited gradients that cannot be replenished. What appears dynamic is merely the rearrangement of a finite stock under terminal conditions. The Hospice Zone is not a cosmos in motion but a managed decline, a controlled descent toward annihilation.

3.1.1 Structural Definition

The Hospice Zone is defined by three non-negotiable conditions:

1. **Finite Energy Gradients:** No external influx of fresh order exists. All dynamics are internal redistributions of a fixed, original stock.
2. **Monotonic Entropy Increase:** The system moves irreversibly toward lower-order, higher-entropy states.
3. **Temporary Temporal Suspension (The Fuse):** Time in the Hospice Zone is not a neutral backdrop or a symmetrical axis. It is the containment protocol that holds the world in suspension until the Absolute Asymmetry of the Infinum reclaims the field. In the medium, time functions as delay with a countdown, friction between intention and manifestation created by the drag of Sin Mass. It serves as a hospice interval, a mercy window that narrows with every passing moment. Time is not a line. It is a fuse, and the fuse is burning. The apparent symmetry of delay is false, because the countdown is not expanding but contracting.
4. **Terminal Processes:** All processes are downstream of decay. None are generative in the ontological sense.

3.1.2 Behavioral Profile of a Decaying Medium

A decaying medium exhibits a predictable set of behavior signatures:

- **Collapse:** Structures lose integrity and fall into compact, lower-energy configurations.
- **Shedding:** Excess momentum, heat, and complexity are expelled as waste.
- **Condensation:** Matter clumps into simpler, more stable forms.
- **Dissipation:** Gradients flatten and distinctions blur.

3.2. Reinterpreting Star Birth as Entropic Condensation

The Hospice Zone cannot generate new entities, and its stellar structures are no exception. What modern cosmology calls star birth is simply the most dramatic expression of entropic condensation within a sealed, decaying medium. Under increasing resistance, matter does not ascend into higher forms of coherence. Matter collapses into denser and more unstable configurations that burn through inherited potential at accelerated rates. The phenomenon is not generative but extractive terminal rearrangement of finite gradients that produces temporary luminosity while deepening the system's overall decay. In a quarantined medium, light is not a sign of creation but the signature of consumption.

3.2.1 The Standard Narrative vs. Forensic Reality

Mainstream cosmology uses the term star birth to preserve the illusion of cosmic fertility. QC accepts the observations but rejects the narrative. Every stage of so-called star formation is a downhill entropic move:

- **Cloud Collapse:** Diffuse complexity contracts into simpler, denser regions.
- **Heating:** Gravitational potential energy degrades into low-grade heat.
- **Shedding (Jets/Outflows):** Waste expulsion allows for further collapse.
- **Stabilization:** A temporary equilibrium is reached as the system radiates energy away.

Stars are not newborns. They are high-intensity decay nodes. A star burns stored potential into radiation, accelerating entropy locally and globally. The star is the residue of collapse, not the product of creation.

3.3 Technology as Death-Energy Extraction

Terrestrial technology mirrors the same decay architecture found in the stars. Innovation is merely the discovery of new architectures for accelerating decomposition.

- **Fossil Fuels:** The extraction of energy from liquefied cemeteries (prior biological decay).
- **Nuclear Fission:** The splitting of unstable nuclei (remnants of prior stellar death).

- **Renewables:** The harvesting of radiation from a collapsing star or the exploitation of pressure differentials in a decaying climate.

Technology does not create energy but releases trapped death-energy from prior decay events.

3.4 Fractal Consistency Across Scales

The **No New Entities** principle is validated by its cross-scale coherence. The same signature of the Hospice Zone appears at every level of the audit.

Scale	**Behavior**	**Forensic Interpretation**
Cosmic	Star formation	Entropic condensation
Planetary	Climate/Geology	Gradient flattening
Technological	Engines/Grids	Death-energy extraction
Civilizational	Growth/Progress	Metabolic consumption
Biological	Metabolism/Cognition	Decay-driven patterning

3.5 Formal Statement of the Principle (Hospice Constraint)

Within a quarantined, decaying medium lacking external input, no fundamentally new entities can be generated. All apparent instances of birth or emergence are reconfigurations that accelerate the medium's overall entropic trajectory.

- **C1: Stellar Corollary:** Stars are entropic condensation nodes.
- **C2: Technological Corollary:** Technologies redistribute decay-derived energy.
- **C3: Civilizational Corollary:** Civilizations are transient metabolic overlays.
- **C4: Narrative Corollary:** Birth and progress are interpretive artifacts used by hospice-bound agents to deny systemic collapse.
- **C5: Temporal Corollary:** Time is the diagnostic of the delay between the Fracture and the Return. It is the finite duration of the quarantine, acting as a buffer that prevents the immediate culmination of decay. The existence of time proves the hospice is terminal. The countdown proves the hospice is intentional.

CHAPTER 4

The Quarantine Event Medium Fracture

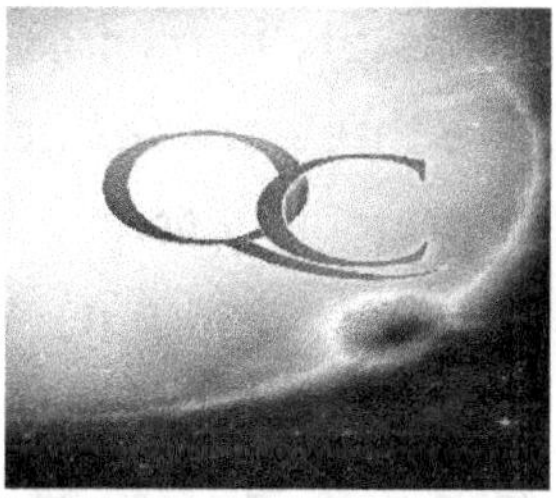

4.0 Introduction

In the DPE framework, the Fall is not treated as a historical episode or a moral narrative. It is a global medium-phase transition that altered the structural condition of reality. The event did not introduce new laws. It introduced Resistance, forcing the original architecture to behave as if it were governed by delay, decay, and distance.

4.1 The Mechanics of the Transition

The transition from the unfallen state Infinum to the quarantined state occurs in three distinct structural phases. This is not a change in the Creator's nature, but a judicial shift in the medium's conductivity. This phase transition represents the exact moment when the singular sufficiency of the Original Architecture was subjected to localized impedance.

Rather than a slow degradation, the transition was a binary state change, a structural snapping of the medium from zero-resistance to a resistive constant. This shift necessitated an immediate reorganization of how energy, light, and presence operate within the affected domain. Because the medium no longer permitted the instantaneous transmission of the divine will, the architecture began to manifest compensatory behaviors while edging towards total systemic collapse. The three phases detailed below identify how the loss of perfect conductivity forced the emergence of the physical world as we currently measure it.

4.1.1 Collapse of Global Coherence

Before the fracture, the medium maintained perfect conductivity. Presence was globally continuous. No phenomenon required propagation. At the moment of rupture, coherence failed simultaneously across the entire medium. The impedance acquisition forced the medium to adopt a non-zero resistance value, causing global presence to collapse into isolated, localized states.

Forensic Note on the Failure of Propagation: The closed-loop framework treats the speed of light as a fundamental constant holy grail of cosmic architecture. In the DPE framework, it is exposed as a terminal hallucination. Light did not begin to move at the Fall. It began to lose coherence. In Infinum, presence is globally continuous, meaning light does not travel because it is already everywhere simultaneously.

The fact that light now possesses a measurable speed is the primary evidence of the Fracture. Propagation is not a feature of a perfect universe. It is the diagnostic symptom of a medium that has acquired impedance. Our current light is slow because it is struggling to traverse the Sin Mass of the Quarantine. To measure the speed of light is not to measure a universal law, but to measure the rate of structural drag within the hospice. Light is the messenger carrying the forensic news that the medium is broken.

4.2 Emergence of Impedance (Sin Mass)

The fracture introduced a new structural quantity: Sin Mass. Sin Mass is not matter, energy, or curvature. It is the Resistance to Coherence. Its introduction makes propagation delay mandatory and makes collapse events, such as particle formation, unavoidable. Sin Mass is the physical cause of symptoms misidentified as fundamental laws. Gravity is revealed as containment pressure, inertia as the resistance to state change, and entropy as the loss of structure over time. Mass itself is nothing more than accumulated impedance.

4.3 Initiation of Time and Distance

Time and distance are not dimensions of reality. They are consequences of resistance. Distance is the integrated impedance between loci of presence. Where resistance is non-zero, traversal becomes a requirement. Time is the ordering protocol required once traversal exists. It is a bookkeeping system for delay. The Formal Logic of the Fall dictates that where there is no resistance, there is no distance, no traversal, and therefore no time. The Fall is the origin of Time.

4.3.1 The Liturgical vs. Chronological Distinction

The celestial bodies established in the original architecture were not created as clocks to measure the passage of duration. In Infinum, where presence is globally continuous, the sun, moon, and stars functioned as Liturgical Beacons. Their purpose was to facilitate cycles of communion and rhythms of worship, such as the Sabbath. These are qualitative states of encounter, not quantitative measurements of delay. The forensic evidence for the non-chronological nature of the Creation Week is two-fold. First, the Seven-Day Outlier has no basis in the astronomical cycles of the Quarantine. It is a structural template of reality. Second, while Genesis discusses years and days as liturgical units, the week is never used as a measurement of duration until after the medium has fractured.

4.3.2 The First Prophecy as the First Tick

Time, as a bookkeeping protocol for impedance, only emerges once the medium acquires Sin Mass. The first true tick of the clock is recorded in the first prophecy of Genesis 3:15. By establishing a sequence involving a promise made and a fulfillment to come, the Auditor initiated the Fuse. Time was not created by God. It was permitted as a mercy-mechanism to limit the duration of sinful existence within a terminal medium.

4.3.3 The Omission of the Week: A Forensic Evidence Point

Chrono-intoxication describes the inability of the observer to perceive reality apart from chronological progression. The state mentioned treats time as a fundamental coordinate of existence rather than a localized constraint of the Fracture. In Default

Eternity, the attributes of God, Deliberate, Perfect, and Eternal, operate without the impedance of decay.

The Seven-Day Week serves as an Architectural Template rather than a clock. The interval of creation did not constitute a 168-hour work week necessitated by the complexity of the task. Instead, the task underwent mapping onto the Seven-Day Template to establish the jurisdictional law of the Sabbath. The template defines the rhythm of the medium, but the act of assembly itself originates from a jurisdiction where the 168-hour limit remains irrelevant.

4.3.4 The Chronological Fallacy of the 168-Hour Window

The literal seven-day creation interval remains a jurisdictional fact, yet it is often obscured by the 168-hour fallacy. Secular cognition, trapped in a state of time-intoxication, assumes that a literal week must be defined by the total duration of 168 hours of elapsed time within the Fracture. This assumption erroneously subjects the Creator to the frictional delay and chronological drag of the quarantined medium.

In the current audit, the seven days are recognized as literal units of organization because they represent the Architectural Template of Default Eternity. The Creator did not wait for the Earth to complete a rotation to define a day. Rather, the rotation of the Earth underwent calibration to reflect the pre-existing Jurisdictional Law of the Template. The literalness of the week is found in the pattern established by the Sovereign Auditor, not in the chronological duration required for the assembly. The act of creation originates from a jurisdiction where the attributes of

God—Deliberate, Perfect, and Eternal—operate without the impedance of decay or the necessity of delay.

4.3.5 The Theological Symbiote and Temporal Projection

The 168-hour fallacy extends beyond secular naturalism into the domain of traditional theology. Many theologians operate under a state of time-intoxication, projecting the chronological constraints of the quarantined medium onto the nature of the Divine. This results in the Theological Symbiote compromise, where the Creator is viewed as a being who exists within time, or worse, as the intentional designer of temporal sequence of decay.

Within the DPE framework, time is not a created dimension but a Diagnostic of Delay. It is the friction generated when the Divine Will encounters the resistance of the Hospice Zone. To treat the Creator as the Architect of Time is to mistake a containment symptom for a foundational design. God did not create time. The Fracture necessitated the chronological gap as a means of managing the terminal decay of the medium.

By reclassifying time as a mechanical result of Impedance, we remove the Boxed-In God from the chronological loop. The Creator remains the Sovereign Auditor of the sequence, standing outside the resistance-born delay that defines our current reality.

4.4 The Axiom of the Count: The Inventory of the Hospice

The transition from Global Coherence to Localization necessitated a forensic change in the Auditor's relationship to the medium. In

Infinum, the concept of a total or a count is a category error. One does not inventory the limitless. However, the observable universe is not Infinum. it is the Quarantine. Scripture records the definitive protocol of the containment zone by stating that He counts the stars and calls them all by name. The declaration establishes the finite nature of the hospice through the finitude of the quarantine. A universe that can be summed is a universe that is bounded. By maintaining a count, the Creator identifies the observable heavens as a finite, bordered environment.

4.4.1 The Summation of the Medium

Because the observable universe is a restricted containment, it can be summed. The act of counting is the logical proof that we are looking at a localized fracture rather than the boundless original architecture. The stars are not scattered across an endless void. They are the specific, numbered inventory of a managed enclosure. To name a star in a forensic context is to assign a unique identifier to a localized point of presence. While the closed-loop framework mistakes the vastness of the Quarantine for infinity, the Auditor recognizes a specific manifest of assets currently held in suspension.

4.4.2 The Intoxication of the Medium

The primary obstacle to recognizing the Quarantine is that the Brilliant Fool is profoundly time-drunk. Having spent his entire existence imbibing the elixir of chronology, he has lost the forensic equilibrium required to see Infinum. He mistakes the swaying of his own distorted perception for the fundamental movement of the cosmos. Because of this chrono-toxicity, he assumes the delay of light is its natural speed and the chronological drag of the

Fracture is the baseline of existence. To the time-drunk observer, a finite, inventoried hospice looks like an infinite, self-sustaining universe. But the Sovereign Auditor remains sober. The Audit is the clinical removal of the chronological haze to reveal the structural blueprint underneath.

4.4.3 The Hubris of the Elite: Reframing Decay as Perfection

The height of fallen hubris is found in the elite effort to reframe the symptoms of the Quarantine as the perfections of a self-sustaining system. The closed-loop framework does not recognize the resistance of the medium. He mislabels it. By taking the measurable drag of what we identify as Sin Mass and the delay of Light and rebranding them as Fundamental Laws, he attempts to lend an air of permanence to a terminal state. He celebrates the balance of a universe that is in a state of managed collapse. The self-praise is the ultimate symptom of the chrono-intoxicated mind. He mistakes the seemingly consistent behavior of a fractured medium for the intentional design of an infinite one.

4.4.4 The Rise of the Chronologian: Theology in the Theological Symbiote

While the Brilliant Fool is blinded by hubris, a secondary error is found in the **Chronologian**. The Chronologian is the theologian who, having become chrono-intoxicated on secular speculative science (time relativity), attempts to trap the Creator within the constraints of the Fracture. By accepting chronology as a baseline reality, he forces the infinite acts of Infinum into the narrow, resistive bottlenecks of the Quarantine, and then forced to create complexity for explanation.

The Chronologian's error is two-fold. First, he treats the delay of the fallen medium and its relative fluctuations as if they were a divine design choice rather than a forensic symptom. Second, he attempts to defend Scripture by making it compatible with the very laws of relativity that the Audit is designed to overturn. In doing so, the Chronologian becomes an unwitting architect of the Theological Symbiote, building a prison of variable time around the One who stands outside the sequence. He fails to see that a God trapped by time relativity is a God trapped by the specific mechanical failures of the Fracture.

UNIT 2: The Resistive Medium (The Physics of the Quarantine)

CHAPTER 5

Sin Mass: The Substance of Resistance

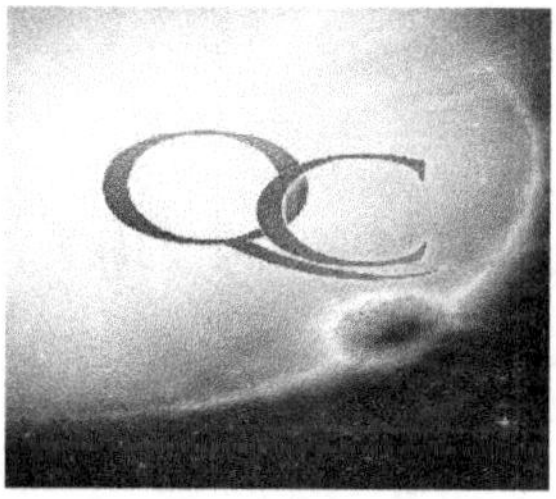

5.1 Definition and Properties

Sin Mass (ρ_{sm}) is the defining physical quantity of the quarantined universe. It is a scalar field representing the density of medium-level resistance to coherence.

Properties of ρ_{sm}:

- **Non-Material:** It is not composed of atoms or subatomic particles.
- **Non-Energetic:** It does not possess work-potential. it hinders it.
- **Accumulative:** Its effects compound over distance and duration.
- **Proportionality:** Directly proportional to the impedance of the medium.

Modern cosmology describes dark matter with three defining traits: 1) outnumbers visible matter roughly 5:1, 2) does not absorb or emit light, and 3) exerts a tethering, binding gravitational effect. These are not incidental but the physical signatures of Sin Mass corruption that has taken on structural weight. The 5:1 ratio proves that corruption is not a surface phenomenon. It is the dominant substrate of the Hospice Zone. The universe is overwhelmingly shaped by what it cannot comprehend.

5.2. Functional Roles of Sin Mass

Sin Mass determines the behavior signatures of the medium in four primary domains. It acts as the operational governor of the Quarantine, translating the absence of absolute coherence into the physical constraints observed by secular science. These roles are not independent laws but are the varied expressions of a single resistive substrate interacting with the remnant energy of Infinum.

5.2.1 Coherence Loss (Particle Generation)

High ρ_{sm} forces global presence to collapse into localized excitations. The collapse is the origin of photons, particles, and the observed wave-particle duality.

5.2.2 Propagation Delay (The Speed of Light)

Propagation speed is not a universal constant of speed, but a throughput limit of a resistive medium.

$$v \propto \frac{1}{\rho_{sm}}$$

As Sin Mass density increases, propagation speed must decrease.

5.2.3 Structural Decay (Entropy)

Entropy is not a fundamental law. It is the dissipation of coherence under resistance.

$$\text{Entropy} \propto \rho_{sm}$$

5.2.4 Containment Pressure (Gravity)

Gravity is the pressure exerted by accumulated Sin Mass.

- **Mass:** Localized ρ_{sm} accumulation.
- **Gravity:** The gradient of ρ_{sm} density.

The gradient creates a tethering effect signature of Sin Mass binding the medium. It drags and creates gravitational debt. It pulls everything toward collapse and it is why galaxies hold together in ways visible matter cannot account for. The universe is literally stitched together by the weight of its own decay.

5.2.5 Darkness as Non-Comprehension

Darkness is failed comprehension, not absence. In the DPE architecture, darkness is the failure of the medium to interpret the Structural Constant (light). Sin Mass is blind to perfection. It cannot receive, reflect, or transmit the DPE orientation. Light travels through it, but the medium does not light up because the substrate is non-responsive. The darkness of space is not emptiness. It is non-comprehension made visible.

5.3 Observable Manifestations

Every major feature of the **Necroverse**—the terminal, systemic environment generated by the friction of the rupture—is a symptom of Sin Mass. Unlike the original creation, the Necroverse is not an expanding or living cosmos; it is a finite theater of decay built upon the **Necrostrate**.

The following table distinguishes the pathological origin from the resulting material foundation of the Hospice Zone to ensure diagnostic accuracy during the Audit.

Table 5.1: Forensic Differentiation of Substrates

Feature	Sin Mass	Necrostrate
Forensic Role	Pathological Agent / Root Cause	Material Foundation / Resulting Anatomy
Ontological Nature	The active principle of resistance; the sludge of non-being.	The corrupted matter and energy in a state of terminal friction.
Operational Function	Generates impedance and drag against the Sovereign's signature.	Serves as the tangible "floor" and substrate for the Necroverse.
Medium Status	The non-being intruder saturating the field.	The scar tissue or forensic remains of the original medium.

The following forensic signatures represent the mechanical failure of the Necrostrate as it attempts to process the Sovereign's signature through a layer of resistance:

- **Redshift:** Decay in transit (loss of energy to the medium), not expansion.
- **Inertia:** The medium's reluctance to change state under resistance.
- **Cosmic Horizon:** The boundary of maximum transmissible coherence.

5.4 Mathematical Behavior

Sin Mass interacts with energy according to the dissipation equation:

$$\frac{dE}{dx} = -\kappa \rho_{sm} E$$

- **Low ρ_{sm}:** Energy remains distributed (Wave behavior).
- **High ρ_{sm}:** Energy localizes (Particle behavior).

Summary

The universe is not expanding or finite by design but resistive. Sin Mass is the physical substance of the fracture, the reason the universe behaves as if it is contained, delayed, and decaying. In the unfallen Infinum, light is not a beam. It is the environment. But in the Hospice Zone, light moves, but the medium cannot interpret it, so the universe remains visually black. The cosmos looks dead because illumination fails even when photons succeed. Infinum will not be dark because there is no Sin Mass and no interpretive distortion. Darkness is a symptom of a world whose substrate cannot comprehend the perfection that surrounds it.

CHAPTER 6

Sin Mass and the Readable Universe: Decay as Diagnostic

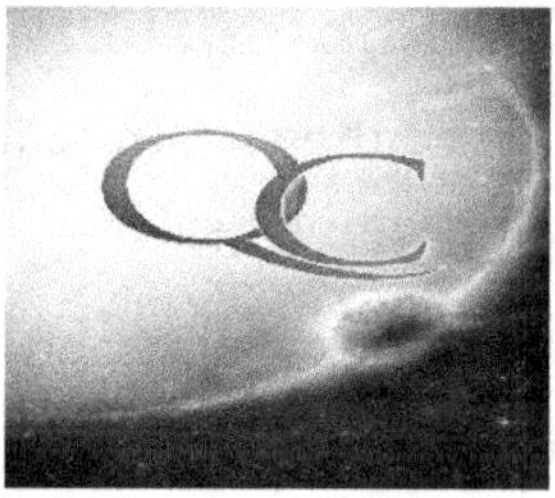

6.0 Introduction

Sin Mass is the foundational contaminant saturating the Hospice Zone. It is not a metaphor, a moral category, or a psychological construct. It is the structural corruption embedded in the substrate of the observable universe.

Once Sin Mass is acknowledged as universal, the cosmos becomes readable. The destructive behaviors of the Necroverse (collapse, decay, exhaustion, and simplification) are no longer viewed as random entropy. They are expressions of a single contaminant acting uniformly across all scales. Paradoxically, the decay is the Good News: it reveals the system's non-originality and points to an external, DPE-defined Creator who sustains the Hospice Zone.

6.1 Sin Mass as the Universal Contaminant

The identification of Sin Mass as a universal contaminant shifts the focus from local anomalies to a global structural failure. In the DPE framework, contamination is not a localized infection but a systemic saturation. It is the fundamental reason the medium no longer mirrors the perfection of Infinum. By analyzing the medium as contaminated, we transition from the mystery of secular physics to the clinical certainty of a forensic audit.

6.1.1 Structural Contamination

Sin Mass is the corruption of the substrate itself. It is functionally embedded in the following domains:

- **Matter:** The inherent instability and eventual decay of the atomic structure.
- **Energy:** The unavoidable dissipation of all gradients through resistance.
- **The Medium:** The impedance that generates the illusion of Time as a linear constraint.
- **The Optical Substrate:** The failure of the medium to interpret the Structural Constant (light), resulting in the physical observation of darkness.
- **The Laws:** The behavior signatures that govern the Hospice Zone, which are misidentified by the closed-loop framework as universal constants.

6.1.2 Legibility Through Contamination

Legibility is achieved when we recognize that the same contaminant produces identical decay signatures across disparate scales. The universe ceases to be an unfathomable mystery and begins to behave as a single system under a terminal constraint.

- **Cosmic Scale:** Stars collapse and galaxies simplify as they exhaust their coherence.
- **Planetary Scale:** Planets cool and ecosystems undergo a progressive exhaustion of vitality.
- **Civilizational Scale:** Structures decay and technological progress serves only to accelerate the rate of entropy.

6.2 Behavioral Signatures of the Contaminant

A contaminated medium cannot produce a diversity of outcomes. It can only produce variations of the same decay behavior. Every destructive behavior in the Hospice Zone falls into one of five diagnostic categories:

1. **Collapse:** Structures lose volumetric and structural integrity over duration.
2. **Shedding:** The mandatory expulsion of waste (heat/radiation) required to facilitate structural collapse.
3. **Condensation:** Matter simplifies into more stable, lower-energy forms as coherence fails.
4. **Dissipation:** The flattening of gradients into a state of thermal equilibrium.
5. **Non-Comprehension:** The inability of the medium to reflect or transmit the DPE orientation, causing light to behave as a localized traveler rather than a native, global condition.

The universe is readable because it is predictable; it is predictable because it is contaminated.

6.3 The Good News: Decay as Evidence of Externality

If the universe were the Original Architecture, it would not exhibit a terminal profile. In the unfallen Infinum, light is the environment. There is no failure of comprehension and therefore no darkness. The blackness of the modern cosmos is not a void of matter, but a void of translation. It is the visual proof that the medium is non-responsive to the perfection of the Creator. Decay is not a flaw in the design. It is proof that it *is not* the design.

6.3.1 The DPE Necessity

The force sustaining a decaying universe cannot be emergent or contaminated. It must be:

- **Deliberate:** Intentional, not a random accident of physics.
- **Perfect:** Uncorrupted, not subject to the medium's decay.
- **Eternal:** Not bound by the Hospice Zone's terminal timeline.

These are not theological adjectives. They are structural necessities. Anything capable of preventing the instantaneous collapse of a contaminated system must be DPE.

The Reproduction Paradox

Naturalistic cosmology insists that decay is normal. The claim is forensically dismantled by the persistence of Human Reproduction.

In a decay-normal universe:

- Biological systems would degrade monotonically.
- Mutation loads would accumulate without limit.
- Extinction would be an immediate and irreversible event.

The Persistence of Information Fidelity

If decay were original, anti-decay behaviors (like cellular fidelity and genetic information integrity) could not exist. The fact that reproduction persists, even in a compromised, error-prone form, proves that the system is being held above the Extinction Threshold by an external input.

Conclusion: Reproduction is not a biological miracle. It is structural proof that the Necroverse is being held together from the outside.

6.4 Formal Diagnostic Loop

Once Sin Mass is acknowledged, the forensic logic becomes unavoidable:

1. **Contamination** leads to **Quarantine**.
2. **Non-Comprehension** (Darkness) confirms the Fracture.
3. **Quarantine** implies **Externality**.

4. **Externality** requires a **Sustainer**.
5. **Sustainer** must be **DPE**.
6. **DPE** points to the **Creator**.

The Necroverse is readable precisely because it is not the source, but the forensic evidence of the Source. This readability extends even to the pervasive darkness of the Hospice Zone where the system appears terminal because it is ontologically blind and suffering from the advanced stages of Chrono-intoxication. Within this medium, Sin Mass acts as the mechanism of transmission failure, ensuring that illumination is neutralized even when photons succeed in their transit across the Necrostrate. The persistent human capacity to recognize this darkness as an abnormality, rather than a natural state, serves as the final forensic proof that the identity was designed for the Absolute Asymmetry of Infinum.

CHAPTER 7

The Necroverse: Deathnology at Cosmic Scale

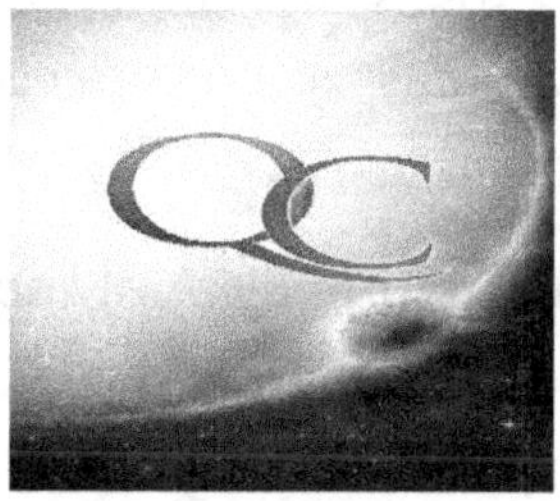

7.0 Introduction

Deathnology was developed to diagnose the structural decay of civilizations, institutions, and cultural systems. Its core insight is that complex structures within a terminal environment follow predictable patterns of exhaustion, simplification, and entropic shedding.

In Quarantine Cosmology, this diagnostic framework is extended beyond the human domain. The same decay architecture that governs societies and technologies governs the Necroverse itself. This extension reveals that the macroscopic structures of the cosmos are not separate from the localized failures of human systems. Both are operational results of the same resistive medium.

- **Cosmological Expression**: The Necroverse is the scaled expression of substrate behavior previously identified at the civilizational level. It is the systemic totality of the Necrostrate in a state of advanced Chrono-toxicity.

- **The Terminal Chamber**: The observable environment is not a generative system, but a terminal decay chamber. Every apparent act of creation or evolution within this theater is merely the reconfiguration of decomposition and the shifting of Sin Mass into more complex, yet equally doomed, arrangements.
- **Predictable Exhaustion**: By applying Deathnology to the cosmos, we recognize that the laws of physics are actually the metabolic rates of a dying system. The exhaustion of a star and the collapse of an institution are forensically identical. Both are shedding energy that can no longer be sustained by the compromised medium.

7.1 From Deathnology to Cosmology: Structural Continuity

Deathnology identifies three universal decay behaviors that appear at every scale of the Hospice Zone:

- **Collapse** of complex structures.
- **Shedding** of excess energy, momentum, and information.
- **Condensation** into simpler, more stable forms.

These behaviors are not limited to civilizations. They appear in biological metabolism, technological systems, and geological processes. QC extends the continuity upward. The universe exhibits the same decay signatures as any hospice-bound system. Cosmology naturally falls into the architecture of Deathnology.

7.2 The Necroverse Defined

The Necroverse is a quarantined, decaying universe in which all processes (stellar, planetary, biological, and technological) are expressions of decomposition. It is governed strictly by the No New Entities Principle:

- **No new** ontological categories can appear.
- **All births** are collapse residues.
- **All innovations** are rearrangements.
- **All growth** is metabolic consumption of finite gradients.

7.3 Stellar Processes as Necrotic Behavior

What mainstream cosmology frames as star birth is a necrotic condensation event.

- **Star Formation:** Molecular clouds collapse, gravitational potential degrades into heat, and matter simplifies as waste is expelled via jets.
- **Stars as Decay Engines:** A star is a high-temperature decay node. It burns stored potential to accelerate entropy.

The Necroverse treats stars as necrotic furnaces, not newborn entities. They are flare-ups of decomposition within a terminal ward.

7.4 Technology and Civilization as Necroverse Behaviors

Technology is the engineering of **decay extraction**. No new energy is created, only death-energy is accessed.

- **Fossil Fuels:** Ancient biological death.
- **Nuclear Fission:** Remnants of stellar death.
- **Solar:** Harvesting the radiation of stellar decomposition.

Civilization is a metabolic overlay powered by the decay. It is maintained by gradient consumption and stabilized by the narrative denial of its terminal status. Civilizations are accelerators of the Necroverse's trajectory toward simplification.

7.5 The Fractal Signature of the Necroverse

The fractal repetition of decay is the diagnostic signature of the Hospice Zone.

Table 7.5: Scale-Invariant Decay Signatures

Scale	Behavior	Necroverse Interpretation
Cosmic	Star formation	Collapse residue
Galactic	Structure formation	Gravitational exhaustion
Planetary	Climate / Geology	Gradient flattening
Biological	Metabolism	Controlled decomposition
Technological	Engines / Grids	Death-energy extraction
Civilizational	Growth / Progress	Metabolic consumption
Kinetic	Light Propagation	Impedance Saturation

Forensic Note: The Metabolic Fallacy—Mainstream biology defines metabolism as a generative process of life-maintenance. Under the Necroverse Audit, metabolism is reclassified as Controlled Decomposition. Biological life does not create energy.

It manages a slowing-down of the decay gradient. A living organism is a localized zone where the medium's inherent collapse is metered and regulated. Death is not a new event that occurs at the end of life. It is the moment the controller (the biological governor) fails, allowing the decomposition to return to the standard, unmetered rate of the surrounding Sin Mass.

7.6 Implications for Cosmology: The End of Generative Narratives

The recognition of the Necroverse eliminates the myths of cosmic fertility, universal creativity, and progress metaphysics. These are narrative artifacts used by hospice-bound agents to avoid the structural reality of the Quarantine. The universe is not a cradle. It is a terminal ward. Its complexity is not growth. It is the turbulence of decay.

7.7 The Kinetic Limit: Light Speed as Necrotic Impedance

The most profound diagnostic of the Necroverse is the existence of a speed limit. In the unfallen Infinum, presence is immediate ($T\_r = 0$). In the Necroverse, presence is replaced by propagation, which is the velocity of a signal fighting through a resistive substrate.

7.7.1 The Speed of Light (c) as a Throughput Limit

The Speed of Light (c) is not a fundamental constant of a generative universe. It is the maximum throughput permitted by a contaminated medium. It is the saturation point of the Sin Mass. Just as a clogged pipe has a maximum flow rate, the Necroverse has a maximum conduction rate for the Structural Constant. The existence of a limit at all is proof that the vacuum is not empty but is a high-resistance quarantine zone.

7.7.2 Time Dilation as Forensic Drag

What mainstream physics misidentifies as time dilation is Kinetic Throttling. It is the mechanical tax imposed by the medium on the state of displacement. Because Infinum is a state of immediate presence, the Necroverse is defined by its inherent incoherency. It is a fragmentation of that presence into the forensic artifacts of distance and travel.

- **The Failed Restoration**: Velocity is the attempt of a fractured medium to achieve coherency through movement. It is an effort to close the gap created by the rupture.
- **The Sin Mass Tax**: This attempt is perpetually hampered by Sin Mass resistance. As an entity increases its velocity to overcome the displacement, the Necrostrate responds with Structural Drag.
- **Impedance Delay**: The resulting slowdown is not a "warping of time," but a clinical accounting of the friction. The more an object forces its way through the substrate, the more the Sin Mass accumulates, ensuring that coherency is never reclaimed.

The Speed of Light (c) remains the ultimate diagnostic of the Quarantine. It represents the maximum rate at which this incoherent medium can attempt to process the Sovereign's signature. The existence of travel itself is the proof of our isolation; we live in a terminal ward where the very necessity of moving from one point to another is the primary evidence of our ontological fracture from Infinum.

CHAPTER 8

Light Under Quarantine: Propagation as Evidence of Resistance

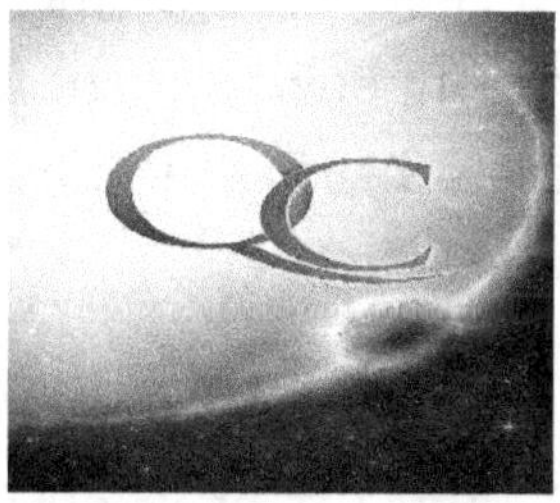

8.0 Introduction

Light is not a fundamental particle, a dual-state excitation, or a universal constant. In Quarantine Cosmology, light is the behavior of coherence under resistance. Its propagation is not a property of reality, but a symptom of the medium's fallen condition.

In the unfallen state Infinum, light does not travel. It remains globally present because coherence is unopposed. Propagation begins only when the medium acquires impedance ($\rho_{sm} > 0$).

8.1 Light as Collapsed Presence

In the pre-fall state, presence is continuous. After the fracture, coherence collapses into localized excitations that must traverse the medium to maintain continuity. A photon is coherence forced into motion. The motion is a forensic marker of displacement. In the healthy architecture, light is the native environment. In the Quarantine, light is a traveler. As a traveler, light is subject to the friction of the medium. All electromagnetic (EM) waves in the Quarantine are compromised and decaying signals. It is why light in the Hospice Zone possesses a dual nature of utility and lethality. Why it facilitates photosynthesis in plants while simultaneously possessing the capacity to induce cellular mutation (cancer) in humans. It is a degraded energy source that carries the tax of the Sin Mass. The transition from environment to traveler is the primary evidence that the substrate has become non-responsive to the Structural Constant.

- **Infinum:** Coherence + Zero Resistance → Global Presence
- **Quarantine:** Coherence + Resistance → Propagation

8.2 The Speed of Light as Maximum Throughput

The speed of light (c) is not a universal constant of speed. It is the maximum throughput of a resistive medium.

$$c \propto \frac{1}{\rho_{sm}}$$

As resistance (ρ_{sm}) increases, throughput (c) decreases. The speed of light is not a property of light itself but strictly a property of the Quarantine Barrier.

8.3 Wave-Particle Duality as Instability

Wave-particle duality is not paradoxical. It is the structural instability produced when coherence interacts with a resistive medium.

- **Wave behavior:** Occurs when coherence remains distributed across the medium.
- **Particle behavior:** Occurs when coherence collapses under local impedance.

Duality is a structural fault in the medium's architecture, not a fundamental feature of reality. The fault extends to the observer's perception. Because the medium is saturated with Sin Mass, it cannot comprehend or translate the fullness of the DPE orientation. Duality is the flicker of a system trying to interpret a signal it is no longer equipped to receive.

8.4 Redshift as Decay in Transit

Redshift is not evidence of cosmic expansion (The Big Bang Myth). It is the loss of coherence during traversal. As an excitation moves through the medium, it loses energy to the resistance of Sin Mass:

$$\frac{dE}{dx} = -\kappa \rho_{sm} E$$

The resulting shift toward lower frequencies is a diagnostic of medium density, not recession velocity.

8.5 The Cosmic Horizon as a Resistance Boundary

The observable universe is not limited by age or expansion. It is limited by maximum transmissible coherence.

Beyond a specific resistance threshold, light cannot maintain its structure and dissipates into the background medium. The horizon is impedance-defined, not temporal.

The universe, therefore, is not expanding away from us. It is resisting us.

8.6 Darkness as Failed Comprehension

The blackness of the vacuum is the ultimate diagnostic of light under quarantine. Darkness is not the absence of light. It is the failure of the medium to interpret the compromised signal.

- **The Interpretive Gap:** Even though compromised photons traverse the Hospice Zone, the substrate (Sin Mass) remains dark because it is blind to perfection. It cannot receive, reflect, or transmit the DPE signal.
- **Visual Proof of the Fracture:** If the medium were healthy, the universe would be a native environment of illumination. The fact that light travels through a corrupt, resistive medium without illuminating it proves that the substrate is non-responsive. The darkness of space is non-comprehension made visible.

8.7 Summary of Light as a Diagnostic

Light behaves as it does only because the medium is damaged and:

- coherence collapsed
- resistance accumulates
- energy decays
- impedance forces collapse
- coherence cannot transmit indefinitely

In the clinical framework of the Necroverse, compromised light is stripped of its jurisdictional power and rendered incapable of true illumination. Because it exists as an untranslated signal struggling through a blind medium, the resulting void remains ontologically dark even when it is physically saturated with photons. The particle completes its transit across the Necrostrate, yet the illumination itself fails to manifest.

This radiation functions pathogenically within the Hospice Zone by providing the necessary throughput to sustain biological metabolism while simultaneously inducing systemic decay. The presence of this dual nature identifies the light of our current cosmos as a damaged non-original energy form. It stands as a forensic shadow of the immediate presence found in Infinum.

Light is the clearest forensic evidence that the universe is not the Original Architecture. In the Necroverse, light succeeds as a damaged signal but fails as a native environment. The darkness surrounding the stars is the physical weight of the Sin Mass world that can register the traveler but cannot comprehend the Source. It is the final forensic proof that we inhabit a medium where even the illumination is slowly degrading the observer.

Just as staring at the sun causes immediate retinal scarring, existing within this resistive medium causes a slow-motion blinding of the observer's diagnostic capacity. The very energy we use to navigate the Hospice Zone acts as a forensic solvent, wearing down the observer's ability to distinguish between the fractured Sequence and the immediate presence of Infinum. By the time the audit is complete, the observer is as non-responsive as the substrate they inhabit.

CHAPTER 9

The Inversion of Need: Psychological Projections of the Necroverse

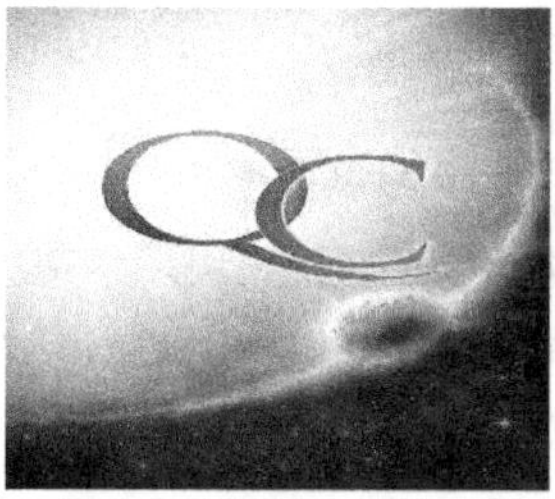

9.0 The Psychological Proof of Inversion

The Inversion of Need provides forensic confirmation of a collapsed hierarchy. The fallen mind, operating within Quarantine cognition, cannot conceive of existence without scarcity, survival, or competition. As a result, it projects these motives onto hypothetical extraterrestrial beings.

The projection is an architectural inevitability under the Ontological Constraint, rather than a creative lapse. When the imagination encounters categories it cannot process, abundance, intrinsic power, non-competitive existence, it substitutes them with distorted analogs drawn from within the Hospice itself.

9.1 The Hierarchy Collapse

If a being has attained high-tier manipulation of Phrysics and gravitational mitigation, it has, by the internal logic of the narrative, bypassed the crude dependencies that would make stealing water, minerals, or genetic material a rational necessity. To suggest that an entity capable of such sophisticated Thanotic technology would traverse the void to harvest the primitive Necrostrate of a hospice world is a forensic contradiction.

Imagining these travelers hunting for terrestrial resources is the forensic equivalent of imagining a high-court judge breaking into a condemned tenement to steal rusted plumbing. It is not a failure of individual creativity but an architectural constraint of the fallen mind. Because the fallen mind remains bound by the decree of Genesis 3:17, secular observation cannot conceive of an existence that does not involve the sorrowful toil of extraction.

9.2 The Theological Inversion: A Parody of Divine Return

The Inversion of Need functions as a precise theological counterfeit. It preserves the drama of higher-power arrival while systematically stripping that arrival of moral weight. Because the fallen imagination cannot utilize holiness, it substitutes it with need. Because it cannot process justice, it replaces it with invasion.

The Original Architecture is inverted at every point:

- Divine abundance becomes alien scarcity.
- Divine judgment becomes alien invasion.
- Angelic hosts become alien fleets.
- Resurrection becomes abduction or experimentation.

What remains is spectacle without authority and contact without consequence.

9.3 Conclusion: The Diagnostic Mirror

Ultimately, the Inversion of Need exposes the final boundary of the fallen imagination. Alien mythology persists not because it explains the unknown, but because it allows the observer to preserve familiar categories (scarcity, fear, and survival) while projecting them safely beyond the horizon of accountability. It functions as a strategic diversion from within, keeping the inhabitant of the Hospice focused on cosmic scraps rather than confronting the source of their condition.

Unit 3: Coherence Topologies (The Architecture of Matter)

CHAPTER 10

Time: The Protocol for Delay

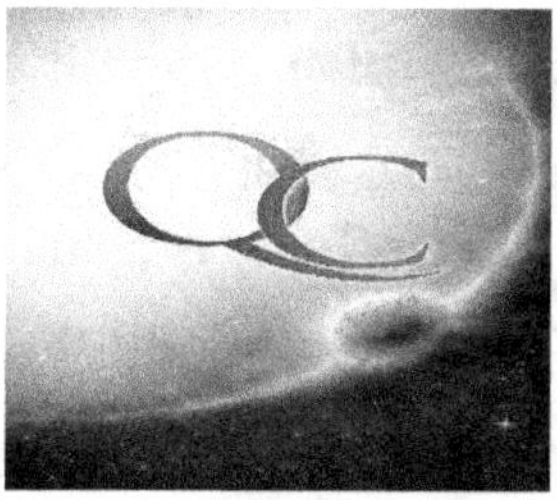

10.0 Introduction

10.0 Introduction

Time is not a dimension of reality, a fundamental fabric, or a flow. In Quarantine Cosmology, time is an accounting protocol necessitated by the transition from global coherence to localized propagation. It is the bookkeeping system for Delay.

In the unfallen state Infinum, time is non-operative because there is no resistance to oppose presence. Time emerges only when the medium fractures and coherence is forced to pay the cost of distance. God, being Deliberate, Perfect, and Eternal, does not ground creation in ontological succession. Succession implies incompletion, movement toward fulfillment, and structural dependency on before-and-after. Such dependency contradicts the fullness implied by perfection. Time, therefore, is not original architecture but a remedial containment concession introduced to prevent the eternalization of corruption.

The Tree of Life: Metabolic Immortality vs. Inherent Dualism

The removal of the Tree of Life demonstrates the principle that sinful immortality is disallowed. Immortality in the Original Architecture is not an inherent property of the soul, as it is commonly understood. Immortality is a sustained metabolic state requiring access to a specific external throughput.

The forensic removal of this access ensures that the rebel cannot eternalize their state of iniquity. If the wicked were to exist in a state of Eternal Conscious Torment, they would require a second, unauthorized immortality independent of the Tree of Life. This would necessitate either an inherent immortality (Dualism) or a Divine maintenance of their suffering (Judgment as Maintenance). Both collapse under the DPE Axiom. By removing the Tree, the Sovereign ensures that the sequence of sin must reach a terminal conclusion. Temporal limitation, therefore, becomes the ultimate mercy.

Time is the structural boundary imposed because of sin. Time is the measurable form of privation, the signature of rupture within a medium that was meant for eternal presence.

10.1 The Origin of Sequence

Sequence is a diagnostic signature of a failed medium. In a perfectly conductive environment, all points of intent are immediately present.

- **Pre-Fall:** Absolute Simultaneity (no ordering required).

- **Post-Fall:** Mandatory Traversal (ordering required to prevent structural overlap/chaos).

Time is the waiting room created by the medium's resistance. We do not move through time. we experience the **latency** of the medium.

10.2 Time as a Function of Resistance

Time (t) is directly coupled to **Distance** (D) and **Sin Mass** (ρ_{sm}). Because distance is the integral of resistance, time is the measure of the duration required to overcome that resistance.

$$t = \int \frac{1}{v(\rho_{sm})} \, dD$$

10.2.1 The Category Error: Clock Dilation vs. Time Dilation

It is a fundamental error of secular cosmology to assume that Time itself dilates or stretches. Within the Quarantine, what is observed is not the warping of a temporal dimension, but the Impedance Throttling of physical systems.

- **The Mechanism:** Every physical apparatus (atomic clocks, biological cells, mechanical gears) is a localized system of coherence operating within the resistive medium.
- **The Interaction:** When a system moves at high velocity or enters a high-density ρ_{sm} gradient, the medium's resistance to the system's internal coherence increases.
- **The Result:** The system's internal processes are physically inhibited. The clock slows because the medium is thicker, not because time has changed.

Forensic Conclusion: Time dilation is a category error. We observe Clock Dilation as the physical manifestation of increased impedance on a localized mechanism. A clock slowing down in high ρ_{sm} is no more time travel than a car slowing down in deep mud is distance dilation.

10.3 The Rejection of the Void (The Deliberate Constraint)

The Deliberate attribute of the DPE Axiom precludes the existence of Empty Space or a Void. God does not create waste, gaps, or ontological nothingness.

- **Secular View:** Space is a vacuum container—a void of nothingness where matter occasionally appears as discrete objects.
- **QC View:** The medium is a saturated field of intent currently in a state of fractured impedance. What appears to be empty space is the Structural Presence of Sin Mass.

The perception of a void is a diagnostic failure of the observer. In the Necroverse, intent has been forced into extreme Localized Compression, leaving the surrounding medium in a state of un-collapsed resistance. This seemingly empty region is the Necrostrate itself. It is a dense field of Saturated Incoherence.

This incoherence provides the observable universe with the illusion of vast scale and expansion. Rather than the creation of new space, expansion is the forensic signature of an unraveling medium. Because the original coherence has collapsed, the medium exhibits an elastic failure. This is a drift of the localized compressions as they lose their architectural grip.

But this expansion is not infinite. It is subject to the Sin Mass Tether. The medium can stretch within its incoherence, but the Quarantine Barrier prevents the contagion of this drift from expanding beyond the designated terminal zone. The void is not empty. It is a dense, non-conductive barrier of Sin Mass that imposes the mechanical cost of Distance and Time.

10.4 Time as a Persistence Protocol

Because the Necroverse is a terminal environment, the Hospice Zone, its contents should collapse instantly. Time functions as a Persistence Protocol, slowing the rate of decay so the restoration process can unfold within a controlled interval.

- **Time is a Buffer:** It prevents the instantaneous collapse of the fractured medium.
- **Time is Temporary:** When resistance is removed in the Final State, the protocol dissolves because it is no longer required.
- **Time is a Terminal Guardrail**: By decoupling the observer from the Tree of Life and subjecting them to the Time Protocol, the Creator ensures that the blinding effect of the Necrostrate (established in Chapter 8) cannot persist forever. The clock is the proof that the suffering has an expiration date.

Summary:

- **Time is not a dimension:** It is a measurement of delay.
- **Time is not fundamental:** It is a symptom of Sin Mass.
- **Time is not a void-filler:** It is the accounting of the medium's impedance.

Time is the primary evidence that the universe is currently operating at a sub-optimal, resistive velocity.

10.5 The Forensic Deconstruction of Temporal Paradoxes

Because Time is a protocol for delay and not a dimension, the paradoxes that plague secular physics vanish when viewed through the lens of Impedance Throttling.

10.5.1 The Twin Paradox as Cosmic Jet Lag

In secular cosmology, the Twin Paradox suggests a literal warping of time that preserves the traveler. In Quarantine Cosmology, the traveler's metabolic and atomic clocks are not preserved. They are battered by the medium.

- **The Mechanism:** The traveler forces their internal systems through the high-impedance medium at extreme velocity.
- **The Interaction:** Velocity increases the rate of collision with the resistive incoherence of the Sin Mass. This creates a state of intense frictional drag and radiation exposure at the molecular and cellular level.

The Clinical Logic of High-Velocity Attrition

This Cosmic Jet Lag is the physical manifestation of movement within a corrupted substrate. In a resistive medium, higher velocity results in a higher entropy load.

- **Environmental Stress:** High velocity equals a higher frequency of interaction with the pathogenic nature of the Necrostrate per unit interval.
- **Accelerated Attrition:** Just as a vehicle wears out faster when driven at high speeds through deep mud, the human

architecture wears out faster when pushed against the Kinetic Throttling of the fracture.

- **The Inverse of Relativity:** While the Brilliant Fool imagines a fountain of youth in the stars, the forensic reality is that movement through the fracture is an agent of exhaustion.
- **The Forensic Verdict:** The traveler returns older and more degraded than the stationary observer. They did not skip time. They accumulated a massive Entropy Load. Their internal decay was accelerated by the thickness of the quarantine.

10.5.2 The Involuntary Admission of High-Velocity Stress

Even the internal diagnostics of secular space agencies have inadvertently confirmed this entropy load. When subjecting biological architecture to high-velocity orbital transit, the result was not the preservation of the traveler. The data revealed systemic genomic instability, chronic inflammation, and cellular fatigue.

The secular observer misinterprets this as a series of unrelated health risks to be solved by better shielding. Within the clinical framework of the Necroverse, these are not risks. They are the mechanical evidence of the medium itself. The traveler is being processed by the friction of the Sin Mass. These biological signatures are the forensic proof that the speed of the traveler is an accelerator of decay and thus aging.

10.5.3 The Illusion of Spacetime Curvature

General Relativity interprets the slowing of clocks near massive bodies as the warping of spacetime. QC identifies this as a High-Density Impedance Gradient.

- Massive bodies are high-density accumulations of Localized Compression.
- The field surrounding these bodies is a gradient of saturated Sin Mass.
- **The Forensic Reality**: A clock near a planet or star slows down for the same reason a light wave redshifts. It is operating in a more resistive region of the medium. The oscillator must perform more work to overcome the density of the Sin Mass. This increased workload causes a mechanical lag in the frequency. The clock slows because the medium is physically obstructing the completion of the cycle.

10.5.4 Velocity as a Systemic Stressor

The secular observation that a clock on a high-velocity aircraft slows down is often used to prove that time is a flexible dimension. Quarantine Cosmology corrects this diagnostic error by identifying the physical cost of traversal.

- **The Physical Struggle**: Movement is the act of forcing a localized system through the Sin Mass. At higher velocities, the system encounters more resistance per unit interval.
- **The Clock as a Stress Gauge**: The clock on the plane slows down because the energy required to maintain its internal

frequency is being diverted or resisted by the friction of the medium.

- **Frictional Lag**: Just as a person running through water moves slower than a person running through air, the internal machinery of a clock (atomic or mechanical) lags when forced through the high-impedance Necrostrate at speed.

The slowing of the clock is the proof of the stress. If the clock is slowing down, the biological and structural integrity of the traveler is also being ground down by that same friction. There is no preservation in speed; there is only the acceleration of wear.

Forensic Conclusion: The Chernobyl of the Stars

The secular ambition for interstellar travel serves as a diagnostic masterclass in deception priming. By framing the high-impedance medium as a frontier to be conquered, the brilliant fool ignores the physical reality of the Necrostrate. High-velocity traversal of light-years constitutes a sprint through a cosmic Chernobyl.

In a resistive medium saturated with sin mass, velocity generates a continuous collision with attrition instead of offering a sanctuary from time. The traveler undergoes a process of accelerated frictional wear within the quarantine rather than a journey of exploration. Any biological architecture forced toward the impedance limit (c) resolves into structural failure long before reaching a destination. The mechanical impossibility of such a mission under the current protocol reveals the dream of space colonization as a high-speed oven designed by deathnology.

10.6 Time as a Legal and Mercy Protocol

If the medium is a Hospice Zone, the delay is not an accident of physics. it is a requirement of Restoration.

10.6.1 The Throttling of Decay

In a fractured medium without the Time Protocol, the collapse into entropy would be instantaneous. By imposing a Maximum Throughput Limit (c), the Creator prevents the immediate dissolution of the system.

- Time is the brake pedal applied to a car heading toward a cliff.
- The Delay provides the Legal Space required for the inhabitants of the Quarantine to experience the choice between coherence (restoration) and total collapse.

10.6.2 Prophecy: Access Without Latency

Prophecy is traditionally misinterpreted as the prediction of a destination that does not yet exist. Within the DPE framework, it is recognized as **Latency-Free Observation**.

- **The Perspective of the Source:** Because the Creator exists in Infinum (Default Eternity), the Sin Mass density is zero ($\rho_{sm} = 0$). In this state, there is no delay to account for. Intent and resolution are simultaneous.
- **The Definition of the Future:** What the quarantined observer calls the future is the unresolved resolution of an intent that is currently traversing the resistive medium. It is future only because the medium's impedance has not yet permitted it to arrive.

- **The Mechanism of the Prophet:** A prophet is not a time traveler. They are a witness granted a temporary, structural bypass of the medium's impedance. They are shown the resolution of the intent as it exists in the Original Architecture, prior to its throttling by the Quarantine.

Forensic Link: It explains why prophecy is often described in Scripture as seeing things that are not as though they were. They are not yet here in the resistive medium, but they are already resolved in the Mind of the Source

10.6.3 The Sentinel Drainage: Black Holes as Impedance Locks

The secular interpretation of black holes as gravitational singularities or gateways is a diagnostic failure. Within the clinical framework of the necroverse, these high-density nodes function as the structural sinks of the quarantine. These sentinels act as gravitational locks, preventing the contagion of sin mass from escaping the localized containment field.

Black holes are commonly thought of as a mathematical singularity or a hole in space. But they are a **Maximum Containment Zone** where sin mass approaches saturation. At this saturation point, the mechanical properties of the medium shift:

- **Coherence cannot escape**: The medium is too thick for transmission.
- **Propagation halts**: The throughput (T) drops to zero.
- **Time/Distance**: Accounting becomes impossible as delay becomes infinite.

The event horizon is not a point of no return in a journey but the boundary where the Sin's Breeze becomes a wall. It is the specific radius where the inward pressure of Sin's Breeze achieves such high density that throughput falls below the threshold required for coherence to transmit.

The secular claim of infinite density is a forensic error that mistakes a mathematical breakdown for a physical attribute. **This mathematical limit** is merely the signal that the Necrostrate has reached absolute elastic failure. It is a mathematical becoming that indicates the localized architecture is undergoing a terminal compaction it cannot survive. These nodes represent the terminal processing of incoherence not wishful escape hatches to other worlds. The black hole is the forensic proof of a closed-loop containment system, ensuring that the corrupted medium remains localized and cannot interfere with the original architecture of Infinum.

10.7 The Mechanics of Clock Dilation (Impedance Throttling)

In the Necroverse, a clock is not a sensor measuring a dimension. it is a localized oscillator fighting against a resistive medium.

10.7.1 The Oscillator-Medium Interaction

Every time-keeping mechanism requires the movement of energy through the medium.

- **High ρ_{sm} (Gravity/Velocity):** The density of the medium increases the drag on internal oscillation.

- **Result:** The system requires more energy to complete a single cycle, or the cycle itself takes longer to resolve. It is not a change in the flow of time. It is a change in the operational efficiency of the clock.
- **The Forensic Nail**: This slowed frequency is the diagnostic proof that the system is performing work against a resistive medium. Within the DPE framework, resistance necessitates friction, and friction necessitates wear. The slowed clock is not a sign of temporal stretching but the stress gauge of the Necroverse. The more the clock lags, the more the internal architecture is being subjected to frictional attrition.

10.8 The End of the Protocol

As a protocol, Time has a Term of Service. It is not an eternal feature of the universe.

- **The Termination Event:** When the Sin Mass is removed ($\rho_{sm} \to 0$), the integrated cost of distance vanishes.
- **The Result:** The accounting protocol (Time) becomes mathematically undefined and structurally unnecessary.
- **The Transition:** Reality returns to Absolute Simultaneity, where intent and presence are once again globally continuous.

10.9 Summary Table: Diagnostics

Feature	Forensic Reality	Secular Error
Time	Accounting protocol for Delay.	A fourth dimension of reality.
Space	Saturated field of Sin Mass.	An empty vacuum.
Clock Dilation	Measured Impedance Lag.	Time stretching or warping.
Velocity	Entropy Accelerator (Friction).	A means of preservation.
Expansion	Elastic failure of the medium.	Dark Energy / Inflation.
Future	Unresolved resolution of intent.	A destination that does not exist.

CHAPTER 11

Gravity: Containment Pressure (Sin's Breeze)

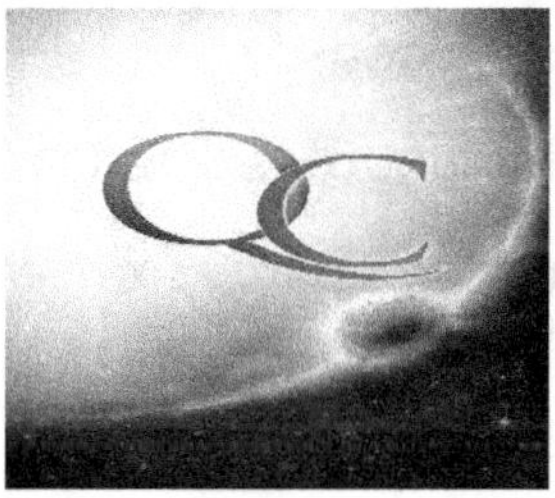

11.0 Introduction

Gravity (Sin's Breeze) is not a force, a field, or a geometric deformation of spacetime. In Quarantine Cosmology, gravity is the containment pressure generated by gradients in Sin Mass. It is the structural response of a resistive environment attempting to contain coherence that can no longer remain globally continuous.

Where Sin Mass accumulates, the medium compresses. Where the medium compresses, coherence is forced inward. The inward pressure of sins breeze is the clinical manifestation of the medium pressing against the density of sin mass, functioning as an entropic push that maintains the containment tether of the quarantine.

11.1 Definition and Formula

Gravity (G) is the gradient of Sin Mass density (ρ_{sm}) across the medium. It is the spatial measurement of how thick the breeze of resistance is at any given coordinate.

$$G = \nabla \rho_{sm}$$

- $\boldsymbol{\rho_{sm}}$**:** Local Sin Mass density.
- $\boldsymbol{\nabla\rho_{sm}}$**:** Spatial gradient of resistance.

The definition removes mass-energy from the foundation of gravity. Mass is not the cause of gravity but the result of accumulated resistance.

11.2 Mass as Localized Resistance

In Quarantine Cosmology, matter is not a fundamental substance but an emergent property of a damaged medium.

- **Mass:** Localized ρ_{sm} saturation.
- **Inertia:** The inherent resistance to a change in state within a thick medium.
- **Gravity (Sin's Breeze):** The pressure exerted by these resistance gradients.

Heavier objects contain more localized resistance, not a matter of pulling more, creating a steeper pressure gradient in the surrounding medium. Inertial mass and gravitational mass are identical because both are different measurements of the same variable: ρ_{sm}.

11.3 Pressure, Not Pull

Gravity does not pull objects together. The medium pushes them toward regions of higher resistance. The Push (Sin's Breeze) explains several phenomena that secular physics struggles to unify:

- **Universal Fall Rate:** Pressure equalizes independently of an object's composition.
- **Weakness of Gravity:** Containment pressure is a secondary effect of resistance, not a fundamental primary interaction.
- **Interaction with Time:** Time is the accounting of resistance. Gravity is the gradient of that same resistance. They are mathematically linked by the density of the medium.

11.4 Curvature as Strain, Not Mechanism

General Relativity correctly describes the *behavior* of gravity but misidentifies the cause. Curvature is not the origin of gravity. curvature is the **strain pattern** produced by resistance gradients.

- **High ρ_{sm}:** Strong curvature (Severe strain).
- **Low ρ_{sm}:** Weak curvature (Light strain).
- **Zero ρ_{sm}** Infinum : No curvature.

The secular concept of **spacetime geometry** is therefore an emergent map of the medium's damage, not a fundamental fabric of reality.

11.4.1 The Forensic Correction of the Fabric

Within the closed-loop framework spacetime is viewed as a trampoline that bends under a weight. In Quarantine Cosmology, there is no trampoline. There is only a saturated medium (the Anti-Void).

- **The Secular Error:** Believing the bend is a feature of space.
- **The Forensic Reality:** Recognizing the bend is the refraction of intent through a thickened medium.

11.5 Cosmological Diagnostics

- **The Universe is not held together by attraction:** It is held together by containment pressure.
- **Dark Matter:** It is misinterpreted resistance. Unseen mass is simply unmeasured ρ_{sm} in the medium.
- **Dark Energy:** It is misinterpreted release. The apparent acceleration of the universe is the medium relaxing or shifting under uneven resistance gradients.

11.6 Summary

Gravity is not curvature but the containment pressure of Sin's Breeze.

- **Mass** is localized resistance.
- **Inertia** is resistance to change.
- **Gravity** is pressure from gradients.
- **Black Holes** are saturation zones.

Gravity is the primary evidence that the universe is not an open system of attraction, but a closed system of containment. The following case study identifies the specific planetary consequences of the Sovereignty Schism as it manifests within the terrestrial interior.

FORENSIC CASE STUDY #1

Sin Mass Geophysics

Earth as Hospice Zone Ground Zero

The cosmology implies several structural consequences:

- Sin Mass increases drag on coherence.
- Drag produces curvature.
- Curvature produces gravity.
- Gravity produces pressure.
- Pressure produces heat.
- Heat produces decay acceleration.

When this causal chain is applied to a planetary body, Earth's interior becomes a pressure-driven furnace of Sin Mass consequences.

1. Earth as the Deepest Compression Well in the Quarantine

Within the DPE architecture, Sin Mass functions as a structural contaminant that distorts the Spatial Medium. Sin Mass introduces resistance to coherence, forcing the medium into curvature and accelerating decay. Although all mass-bearing structures in the Hospice Zone are affected, Earth represents the most concentrated expression of the rupture. As the origin point of the contaminant's introduction into the medium, Earth occupies the deepest gravitational and entropic well in the quarantined system.

The planet is therefore not a neutral geological object but a structurally compromised sphere whose interior reflects the highest density of coherence collapse in the terrestrial domain.

2. The Collapse Chain and the Planetary Fever

The DPE Axiom defines the original architecture as frictionless and Eternal. Sin Mass disrupts this state by introducing drag on coherence, which produces curvature, curvature produces gravity, gravity produces pressure, pressure produces heat through friction, and heat accelerates decay..

This chain of consequences transforms Earth's interior into a system of escalating entropic stress. The planet's heat is the ongoing thermal signature of coherence being resisted within a corrupted medium, not a remnant of formation. The result is a planetary fever that cannot resolve itself under current conditions.

3. The Core as the Primary Sin Mass Sink

In conventional geophysics, the core's heat is attributed to pressure, radioactive decay, and residual formation energy. In Sin Mass geophysics, the core is the deepest compression well of the contaminant. Sin Mass behaves as a downward vector, concentrating at the lowest gravitational potential.

The core therefore represents:

- the densest Sin Mass zone,
- the region of greatest curvature,
- the most intense entropic furnace within the planet.

Its heat is the direct result of coherence being forced through a medium under maximum distortion.

4. Heat as Entropic Friction

Heat is reinterpreted as the frictional byproduct of coherence encountering resistance. The core's temperature reflects the magnitude of this resistance. The mantle's convection reflects the redistribution of entropic load. The crust's instability reflects the inability of the surface to contain the internal pressure. Thus, the planet's thermal state is diagnostic. It is a symptom of structural collapse rather than a neutral physical parameter.

5. Volcanoes as Regulatory Rupture-Vents

Volcanoes function as regulatory structures analogous to black holes in the cosmic medium. Black holes localize coherence collapse in the Quarantine and prevent runaway distortion. Volcanoes perform the same function on a planetary scale.

Their roles include:

- venting Sin Mass pressure
- releasing entropic buildup
- preventing catastrophic crustal rupture
- providing controlled failure points

Volcanic systems therefore act as Sin Mass pressure valves within the Hospice Zone, stabilizing the crust by allowing regulated discharge of internal stress.

6. Mantle Convection as Sin Mass Circulation

Mantle convection is traditionally attributed to thermal gradients. In Sin Mass geophysics, it represents the circulation of entropic load through the planetary medium.

- Subduction zones transport Sin-Mass-loaded crust downward
- Mid-ocean ridges release new material upward under entropic pressure.
- Hotspots represent vertical plumes where Sin Mass finds a direct escape path.

The mantle operates as a dynamic redistribution system for the contaminant.

7. Earthquakes as Coherence Fracture Events

The crust functions as a stressed membrane stretched over a pressurized interior. Earthquakes represent coherence fracture events rather than purely mechanical slips.

- Fault lines correspond to stress lines in the Spatial Medium.
- Seismic waves propagate as shockwaves of coherence collapse.
- Aftershocks reflect the system's attempt to rebalance under new pressure conditions.

The crust's instability is a direct consequence of the planet's entropic load.

8. Water–Sin Mass Interaction and Planetary Reconfiguration

Water is a coherence-carrying medium. Sin Mass is a coherence-destroying contaminant. When large volumes of water interact with high-density Sin Mass zones, the result is catastrophic structural reconfiguration.

Such interactions produce:

- crustal rupture
- mantle upwelling

- rapid thermal exchange,
- basin collapse
- mountain formation
- oceanic reorganization

A global water Sin Mass interaction would permanently alter the planet's surface and internal stress distribution.

9. Mountains as Displacement Structures

Mountains arise as displacement structures formed when the crust buckles under Sin Mass pressure. They represent hardened ridges created by upward entropic displacement.

Their formation reflects:

- crustal thickening,
- folding under stress,
- vertical displacement driven by internal pressure.

Mountains are therefore structural scars documenting the planet's entropic history.

10. Biospheric Coherence Degradation

The biosphere mirrors the geosphere. Pre-rupture matter possessed high coherence. post-rupture matter exhibits degraded coherence. Nutrition functions as coherence transfer, and the degradation of biological matter reflects the same entropic decline observed in the planet's interior.

Consequences include:

- reduced vitality
- shortened lifespans.

- increased fragility
- systemic biological decay

The biosphere is a surface-level expression of the same structural sickness present in the core.

11. The Planetary Fever and the Limits of Self-Recovery

A quarantined system cannot self-correct and the core will not cool into stability. The mantle will not settle into equilibrium or volcanoes cease their regulatory function. And the crust will not stop fracturing and the biosphere will not regain coherence. The planet's fever is not a healing response but a symptom of irreversible distortion under current conditions. Earth, as Hospice Zone ground zero, cannot recover without external intervention.

12. Summary

Earth is a structurally compromised sphere where the interior functions as a pressure-driven furnace of Sin Mass consequences. From the core (the primary entropic sink) to the crust (a fractured membrane), every geological process from volcanic venting to seismic fracture is a regulatory response to the planetary fever. This diagnostic profile confirms that the Earth is not a self-sustaining geological engine, but a terminal system under maximum entropic load, incapable of recovery without external architectural restoration.

FORENSIC CASE STUDY #2

The Synchronicity of Collapse

Recalibrating the Decay Constant

A primary objection from the Hospice Zone intellect posits that the agreement between various dating methods, such as radiometric and non-radiometric, serves as proof of deep time. In Sin Mass geophysics, this agreement is reinterpreted not as a measure of age, but as the Synchronicity of Collapse.

1. The Uniform Resistance Filter

Decay chains across different isotopes agree because they are processed through a singular jurisdictional medium. In Default Eternity, matter is coherent and frictionless. Upon the introduction of Sin Mass, the entire periodic table was subjected to the same resistive gradient of the Fracture.

Isotopes are different components of a single engine, not independent clocks. If the fuel (the Spatial Medium) is contaminated with Sin Mass, every component of the engine exhibits a proportional change in performance. The agreement between isotopes is proof of the medium's jurisdictional uniformity. They are all decaying in the same room under the same Sin Mass pressure.

2. The Correlation of Symptoms: Radiometric vs. Non-Radiometric Clocks

The correlation between decay-based clocks and non-radiometric markers (such as ice cores or varves) is identified as a

Simultaneous Diagnostic. During the Water–Sin Mass Interaction (The Flood), the sudden injection of entropic load accelerated both the nuclear decay rates and the mechanical deposition of layers.

- **Radiometric Acceleration:** The increase in Sin Mass density during the reconfiguration pulse forced an immediate spike in decay rates.
- **Mechanical Acceleration:** The catastrophic hydrological displacement generated thousands of layers in rapid sequence.

The apparent agreement between these systems is not proof of antiquity, but proof of the catastrophic magnitude of the event. Just as high blood pressure and an elevated temperature correlate during a fever, these dating methods correlate because they are synchronous symptoms of the same structural rupture.

3. The Cosmic Mirror: Supernova Light Curves

Supernova light curves match decay physics precisely because the Entire Quarantine is under a unified jurisdictional law. The light from a distant supernova must travel through the same resistive Spatial Medium that governs the decay of isotopes on Earth.

The light curve is a spectral signature of the medium's resistance. The match occurs because the Missing Operator's system is internally consistent. The simulation of decay is a universal constraint within the Fracture, ensuring that the inhabitant sees a uniform system of death in every direction. This prevents the observer from recognizing the Earth as a unique jurisdictional ground zero.

4. The Law of Decaying Decay

The Hospice Zone assumes decay rates are constants (k). Sin Mass geophysics identifies that decay itself is decaying (dk/dt). Observable constancy over a short duration is merely a localized plateau in a downward entropic curve. The constant is a mathematical idolatry that ignores the jurisdictional shift from the frictionless Original Architecture to the high-drag medium of the Quarantine.

FORENSIC CASE STUDY #3

The Myth of the Static Variable

The secular consensus operates under a profound logical absurdity. It requires Deep Time to function as a magical agent of transformation, yet it insists that the laws governing that transformation (Constants) have remained frozen. Within the DPE Axiom, this is recognized as a structural impossibility. A system that is falling apart cannot have a fixed rate of falling apart, because the very process of decay reconfigures the medium in which that decay occurs.

1. The Category Error of Linear Projection

The Brilliant Fool observes a microscopic window of stability, perhaps 100 years of data, and projects that line backward for billions of years. This is the equivalent of observing a car's deceleration for one second and concluding that its velocity has been identical since the moment of ignition.

In Quarantine Cosmology, we identify this as a failure to account for the Jurisdictional Shift. The stability observed today is not a fundamental law but a localized plateau on a steep downward curve of *Decaying Decay*.

2. Time vs. Catastrophe

Secular geophysics uses Deep Time to explain the existence of mountains, canyons, and fossil graveyards, assuming a constant drip over eons. However, a constant drip in a frictionless medium produces no work, and a constant drip in a high-drag medium produces only uniform erosion.

The dramatic reconfiguration of the Earth's surface—Mountains as Displacement Structures and Basin Collapse—requires High-Energy, Low-Time events. The Water–Sin Mass Interaction (the Flood) provided the necessary structural pulse to reconfigure the planet in months, not eons. The Deep Time narrative is merely a psychological prop used to avoid the Logical Necessity of a catastrophic beginning. This forensic reality is corroborated by the unified testimony of ancient civilizations, whose independent records attest to the planetary reconfiguration in consistent terms of sudden, high-energy catastrophe rather than gradual transition.

3. The Variable Foundation

If the Spatial Medium is the foundation of reality, and that medium is saturated with the structural contaminant of Sin Mass, then the foundation itself is a variable.

- **Resistance is not constant:** As coherence dissipates, drag increases.
- **Heat is not constant:** As Sin Mass concentrates in the core, the Planetary Fever escalates.
- **Decay is not constant:** As the medium's density shifts, the clocks (isotopes) re-calibrate to the new pressure.

4. The Forensic Mask of Constancy

The Missing Operator promotes the idea of Universal Constants to ensure that the inhabitant of the Hospice Zone feels a sense of false permanence. If the decay rate is seen as an eternal constant, then the observer is blinded to the fact that they are living inside a Fracture that is rapidly approaching its Restoration Threshold.

The Constancy of Decay is a mathematical idolatry that ignores the reality of a groaning creation. By asserting that *Decay Decays* (dk/dt), we restore the urgency of the audit. We recognize that the current laws of physics are not the original template, but the rules of a temporary, pressurized cage.

CHAPTER 12

Entropy: Coherence Dissipation Under Resistance

12.0 Introduction

Entropy is not a statistical inevitability or a probabilistic trend toward disorder. In Quarantine Cosmology, entropy is the dissipation of coherence under resistance. It is the structural consequence of Sin Mass acting on any localized excitation within the fallen medium.

Where resistance exists, coherence decays. Where coherence decays, structure dissolves. The dissolution is what is clinically measured as entropy. It is not a law but a symptom of a fractured architecture.

12.1 Definition and Rate

Entropy (S) is the cumulative loss of coherence due to interaction with the resistive medium.

$$\frac{dS}{dt} \propto \rho_{sm}$$

- ρ_{sm}**:** Local Sin Mass density.
- dS/dt**:** Rate of coherence loss.

Entropy is not a measure of randomness, but a measure of Coherence Degradation. As resistance increases, the rate of dissipation increases proportionally.

12.2 The Condition of Zero Entropy

In the pre-fall state Infinum , where $\rho_{sm} = 0$:

- Coherence is global and unopposed.
- No collapse events or localizations occur.
- No propagation is required, so no energy is lost to the medium.

Entropy is zero because nothing exists to oppose coherence. Entropy is a post-rupture artifact.

12.3 Entropy as the Cost of Traversal

Every movement through the medium extracts a toll of coherence. The extraction is proportional to resistance. As an excitation (such as a photon) moves:

$$dE/dx = -\kappa\rho_{sm}E$$

Energy loss is coherence loss. Coherence loss is entropy gain. Under the diagnostic:

- **Redshift** is entropy.
- **Thermalization** is entropy.
- **Decoherence** is entropy.

- **Aging** is entropy.

Entropy is the receipt for coherence spent.

12.4 The Time-Entropy Identity

Time and entropy are not independent dimensions. Both are functions of the medium's impedance (ρ_{sm}).

- **Time** measures the **delay** (the ordering of resistance).
- **Entropy** measures the **decay** (the cost of resistance).

Where resistance is high, time accumulates faster and entropy accumulates faster. Time is the accounting. Entropy is the payment.

12.5 Entropy and Sin's Breeze (Gravity)

Gravity (Sin's Breeze) compresses coherence into regions of high ρ_{sm}. The compression increases local resistance, which in turn accelerates entropy.

- It explains why stars burn out and massive objects radiate.
- **Gravity** is the containment mechanism. **Entropy** is the resulting degradation of the contained coherence.

12.6 Cosmological Implications

- **Heat Death is not Fundamental:** The universe does not tend toward disorder. It tends toward Coherence Exhaustion. While local concentrations of Sin Mass produce a thermal fever, the total system is running toward a state of spent integrity.

- **Expansion as Decay:** Cosmic cooling is not the result of space stretching, but the dissipation of initial architectural energy as coherence is bled out by the medium's resistance. The medium is thinning and losing tension as it exhausts its original coherence.
- **The Arrow of Time:** The arrow is simply the direction of increasing resistance. Time is the measure of the delay created by the Necrostrate as it impedes the flow of coherence.
- **Degradation, not Aging:** The universe is not getting older. It is simply breaking down under the strain of the Quarantine. This manifests as a dual failure: the local heating of compression wells (Planetary Fever) and the global cooling of the surrounding medium (Energy Dissipation).

12.7 Summary

Entropy is the dissipation of coherence under resistance.

- It begins at the **Quarantine Event**.
- It increases with **Sin Mass**.
- It governs the rate of decay and shapes the flow of time.

Entropy is the clearest evidence that the universe is not sustained by equilibrium but is collapsing under the weight of its own resistance.

CHAPTER 13

The Dissolution of the Impedance

13.1 The Termination of the Fuse

The removal of the Quarantine is an architectural restoration rather than a subtractive event. Just as the Fall did not create matter but introduced Resistance (Sin Mass), the Restoration does not create a new heaven and earth from scratch; it is the surgical removal of the Impedance Medium. When the Fuse of Time reaches its terminal point, the Auditor withdraws the permission for delay. The sequence of waiting ends because the medium no longer requires traversal. The First Prophecy (Genesis 3:15) is finally settled, and the sequence of seconds, minutes, and eons is deleted as a redundant bookkeeping system.

13.2 The Evaporation of Sin Mass

The Great Noise and Fervent Heat described in forensic prophecy (2 Peter 3:10) are the structural symptoms of Phase Re-Coherence. When Sin Mass is removed, the resistance holding particles in a localized, collapsed state vanishes. The heat is not a fire of destruction in a chemical sense, but the energy of localization being released as the medium returns to a state of global conductivity. The localized knot of the particle, which the closed-loop framework misidentifies as solid matter, unties itself, returning to the infinite flow of the original architecture.

13.3 The Collapse of Distance and the Death of Relativity

With the removal of the medium's resistance the integrated impedance we call distance ceases to exist. The stars are not moving closer. The requirement to traverse a high-friction medium to reach them is simply deleted. The phenomenon of clock dilation is exposed as a mechanical drag on physical hardware under entropic pressure. Clocks never measured a stretching of time. A change to the clock does not affect time any more than a change to the thermometer changes the temperature. Clocks merely reported the density of the resistance they were forced to inhabit. There is no longer a variable resistance to cause drag on coherence. The concept of relativity is revealed as a localized symptom of hardware failure within the Fracture. The observable universe stops being a series of isolated points separated by voids and returns to being a single continuous Presence.

13.4 The End of the Inventory

Once the medium is restored to Infinum the Axiom of the Count is fulfilled. The names of the stars served as forensic identifiers during the Quarantine. With the restoration they transition back into the harmonies of the original assembly. The Audit is closed because there is no longer a localized loss to track. The Sovereign

Auditor no longer needs to maintain the count of a finite hospice. The medium has been reintegrated into the limitless Infinum.

CHAPTER 14

The Observable Universe: A Structural Boundary

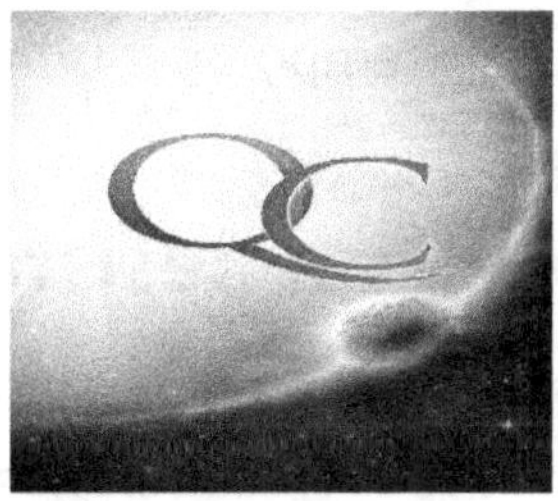

14.0 Introduction

The observable universe is not the totality of reality. It is the bounded region of a resistive medium inside which coherence can still transmit before collapsing. Its size, shape, and behavior are determined by the maximum transmissible coherence allowed by the density of Sin Mass, not by expansion or age. The Observable Quarantined Universe is not expanding. It is contained. It is where everything built so far converges into a single structural picture. It is the moment Quarantine Cosmology stops being a set of isolated insights and becomes a coherent cosmological architecture.

14.1 Definition: The Coherence Limit

The Observable Universe (Ω_{sm}) is defined as the region of the fallen medium within which coherence can propagate without total collapse.

$$\Omega_{sm} = \{x \mid \int_0^x \rho_{sm}(s)\, ds < D_{max}\}$$

Where:

- $\rho_{sm}(s)$ = Sin Mass density along the path
- D_{max} = maximum resistance coherence can traverse

The boundary of Ω_{sm} is not temporal. It is impedance-defined.

14.2 The Horizon as a Resistance Limit

The cosmic horizon is not the distance light has traveled since a beginning. It is the point at which coherence can no longer survive traversal. At the horizon:

- Propagation throughput approaches zero.
- Redshift approaches total energy loss.
- Coherence collapses into noise.
- Information cannot transmit.

The horizon is therefore a containment wall, not an age marker.

14.3 Redshift as a Map of Containment

Redshift increases with distance because resistance accumulates, not because galaxies recede. As light approaches the horizon:

- Coherence decays.
- Frequency drops.
- Structure dissolves.
- Signal becomes indistinguishable from background noise.

It produces the illusion of expansion. What we observe is coherence exhaustion, not recession velocity.

14.4 Large-Scale Structure as Pressure Pattern

Galaxies, filaments, and voids are not the result of gravitational attraction. They are the pressure patterns formed by gradients in Sin Mass.

- **Filaments** form along resistance troughs.
- **Voids** form where resistance is minimal.
- **Clusters** form where resistance accumulates.
- **Superclusters** form where pressure converges.

The universe's structure is not random. It is the medium's stress map.

14.5 Dark Matter and Dark Energy Reinterpreted

The current crisis in standard cosmology, the inability to account for the vast majority of the universe's mass and energy, is the inevitable result of treating the Hospice Zone as a self-sustaining, original system. When the guild observes gravitational effects or expansion rates that defy their localized equations, they invent placeholders to balance the ledger. In the DPE framework, these are not mysterious substances. They are the structural signatures of a medium under judicial constraint. What the observer interprets as missing mass or phantom energy is actually the visible strain pattern of the Quarantine itself.

14.5.1 Dark Matter

Unseen mass is not missing matter. It is **unmeasured resistance**. Regions with high ρ_{sm} behave as if they contain additional mass because they exert greater containment pressure.

14.5.2 Dark Energy

Apparent acceleration is not expansion. It is **medium relaxation**. As resistance redistributes unevenly, coherence decays at different rates, producing the illusion of accelerating recession. Both phenomena are artifacts of containment.

14.6 The Universe as a Quarantine Zone

The observable universe behaves like a containment chamber:

- Bounded by a resistance wall.
- Filled with decaying coherence.
- Governed by delay and dissipation.
- Structured by pressure gradients.
- Trending toward collapse.
- Isolated from Infinum.

The universe is a holding environment.

14.7 Summary

The observable universe is a finite containment boundary defined by the maximum coherence the fallen medium can transmit.

- The horizon is a resistance limit.
- Redshift is decay, not recession.
- Structure is pressure-driven.
- Dark matter is unmeasured resistance.
- Dark energy is medium relaxation.
- Geometry is emergent.
- Time is an accounting protocol.
- Entropy is coherence loss.

The universe behaves exactly as a quarantine zone would behave.

UNIT 4: Bio-Coherence and Transmission (The Human and the Signal)

CHAPTER 15

The Purpose of Containment: Metaphysical Architecture

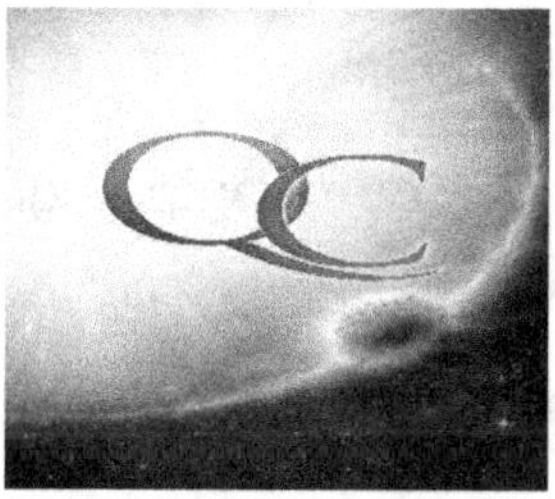

15.0 Introduction

Quarantine Cosmology transitions at the boundary where structural physics meets metaphysical architecture. The quarantined universe does not constitute an experiment, a punishment, or a developmental stage. The cosmos remains a containment environment designed to isolate a corrupted medium from the Original Architecture (Default Eternity or Infinum). The purpose of the environment remains structural.

Containment exists because coherence cannot coexist with resistance. A fractured medium cannot remain integrated with Infinum. Separation is therefore not optional but required for stability. The universe exists in its current state because the alternative is systemic collapse.

15.1 Containment as Structural Necessity

When the medium fractured, resistance became universal. Global coherence became impossible. If the corrupted medium had remained integrated with Infinum:

- Coherence would destabilize.
- Resistance would propagate.
- The original architecture would degrade.

Containment prevents the outcome. The universe is therefore a boundary condition, not a habitat.

15.2 Isolation of Resistance

Sin Mass cannot be allowed to diffuse into the original medium. Containment ensures:

- Resistance remains localized.
- Coherence loss does not propagate.
- Decay is confined.
- Collapse events remain internal.

The universe is a sealed region within a larger coherence field.

15.3 Delay as Stabilization

Delay is not a flaw. Delay is a stabilizing mechanism. Propagation, time, and distance slow the rate at which coherence collapses. Without delay:

- Collapse would be instantaneous.
- Structure would not form.

- The medium would dissolve.

Time is therefore part of the containment protocol.

15.4 Dissipation as Purification

Entropy is not merely decay. It is coherence dissipation that gradually reduces the structural impact of resistance. Over long durations:

- High-impedance regions collapse.
- Coherence exhausts.
- Resistance redistributes.
- The medium trends toward uniformity.

Containment allows corruption to burn itself out without spreading.

15.5 Boundaries as Protection

The cosmic horizon is not a limit of observation. It is a protective wall that prevents:

- Coherence from leaking outward.
- Resistance from leaking inward.
- Collapse from destabilizing the original architecture.

The boundary is a seal.

15.6 Temporary by Design

Containment is not permanent. It persists only as long as resistance persists. When coherence exhaustion reaches completion and resistance falls below threshold, the medium stabilizes. At that point, containment becomes unnecessary. The universe ends because it runs out of resistance.

15.7 Summary

The universe exists in its current state because containment is the only stable configuration for a fractured medium.

- Resistance must be isolated.
- Coherence must be protected.
- Decay must be confined.
- Collapse must be prevented.
- Delay must be enforced.
- Boundaries must be maintained.

Containment is structural. The universe is the safest possible configuration for a damaged medium.

CHAPTER 16

The Restoration Threshold: When Containment Ends

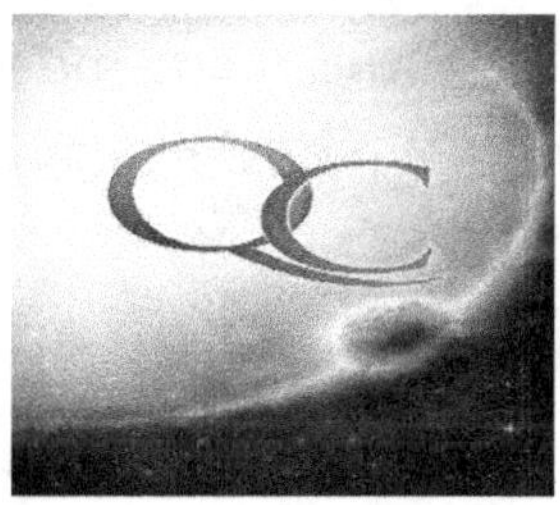

16.0 Introduction

The Restoration Threshold is the structural apex of Quarantine Cosmology, and the moment when the universe no longer requires containment because the medium no longer carries enough resistance to justify delay, distance, or time.

Containment persists only as long as resistance persists. The universe is sustained by the presence of Sin Mass, which forces coherence to propagate, decay, and localize. When resistance falls below a critical threshold, the structural basis for containment disappears. The end of the universe is a condition.

16.1 Definition: The Threshold of Reintegration

The Restoration Threshold (R_t) is the minimum resistance level at which coherence can reintegrate without destabilizing the original architecture.

$$R_t = \{\rho_{sm} \mid \rho_{sm} \to 0 \text{ and } \nabla\rho_{sm} \to 0\}$$

Where:

- ρ_{sm} = Sin Mass density
- $\nabla\rho_{sm}$ = resistance gradient

Containment ends when both the density and the gradient of resistance approach zero. At this point, the Sin Breeze of resistance ceases, and the pressure that holds the Quarantine together vanishes.

16.2 Dissipation Toward Threshold

The universe trends toward the threshold through three specific forensic mechanisms:

- **Entropic Dissipation:** Coherence decays under resistance. As coherence dissipates, the structural impact of resistance diminishes.
- **Gravitational Collapse:** High-impedance regions (localized Sin Mass) collapse inward, reducing gradients. Black holes eventually evaporate as coherence exhausts.
- **Medium Relaxation:** Over long durations, resistance redistributes toward uniformity. Gradients flatten, and containment pressure (Sin's Breeze) decreases.

These processes are the medium's natural trajectory toward stabilization.

16.3 Structural Transition at the Threshold

When resistance approaches zero, the physical properties that define the Necroverse collapse:

- **Propagation:** Becomes unnecessary.
- **Delay:** Collapses.
- **Time:** Ceases to accumulate.
- **Distance:** Becomes non-operative.
- **Curvature:** Relaxes.
- **Entropy:** Halts.
- **Coherence:** Becomes continuous.

Instead of exploding, collapsing, or fading, the universe loses the conditions that make it a universe. The medium transitions from a resistive state back toward Infinum.

16.4 The End of Accounting (No Temporal End)

Because time is an accounting protocol for resistance, the point at which the protocol is no longer required. When resistance falls below the threshold, traversal is unnecessary and delay is zero. Time has no function.The observable universe ends with time.

16.5 The Dissolution of Locality (No Spatial End)

Because distance is accumulated resistance, it is the point at which resistance no longer accumulates during traversal.

- Distance collapses.
- Locality dissolves.
- The horizon becomes irrelevant.

The universe de-localizes.

16.6 Reintegration With Infinum

Once resistance falls below the restoration threshold, the medium can reintegrate with the original architecture without destabilizing it. Reintegration is the removal of the barrier that made separation necessary. The Quarantine ends because containment is no longer required.

16.7 Summary

The universe ends when resistance ends.

- Sin Mass dissipates.
- Gradients flatten.
- Coherence stabilizes.
- Propagation ceases.
- Time halts.
- Distance dissolves.
- Curvature relaxes.
- The horizon becomes irrelevant.

The Restoration Threshold is completion. The universe is temporary because resistance is temporary.

CHAPTER 17

Human Perception Inside a Resistive Medium

17.0 Introduction

Human perception is a local coherence process operating inside a resistive medium. Because the mind itself is constructed from localized coherence, it interprets the properties of the fallen medium (time, distance, decay, and locality) as if they were fundamental. Perception reveals the structure of containment. It is why consciousness experiences the fallen medium as normal.

17.1 Localized Coherence and the Illusion of Motion

In Infinum, coherence is global. In the fallen medium, coherence is localized. Human consciousness is built from localized excitations, propagating signals, and delay-dependent processing. Because a system built from localized coherence cannot perceive global coherence, it can only perceive propagation.

It is the only mode available inside the medium. It is why humans experience:

- **Motion** instead of Presence.
- **Sequence** instead of Simultaneity.
- **Locality** instead of Continuity.

Perception is shaped by resistance. We see the limitations the medium imposes on our cognitive throughput.

17.2 Time as a Cognitive Delay

Neural activity is propagation-based. Every thought is a traversal event and every perception is a delay-dependent sequence. Because the mind is built from delay:

- Time feels intrinsic.
- Sequence feels necessary.
- Causality feels directional.
- Simultaneity feels impossible.

The mind experiences time because the mind requires time to function. The mind's operational latency is fundamental, not time.

17.3 Distance as a Map of Resistance

The brain constructs spatial maps from propagation delays and resistance gradients. Because perception is spatially localized:

- Distance feels like separation.
- Motion feels like traversal.
- Geometry feels objective.

Distance is accumulated resistance. The mind interprets that resistance as space.

17.4 Decay and the Entropic Mind

Human biology is a coherence-limited system. The biology ages because it is constructed from high-impedance matter and continuous coherence loss. Because the mind is embedded in a decaying structure:

- Entropy feels natural.
- Aging feels intrinsic.
- Death feels inevitable.

Entropy is a symptom of containment that the mind has normalized because it lacks a non-resistive reference point.

17.5 The Barrier of Locality

Entanglement is residual coherence, but the mind cannot access it directly because neural processes are local and biological coherence is low. Consequently:

- Non-locality feels impossible.
- Global presence feels absurd.
- Instantaneous coherence feels supernatural.

The mind is not wrong, but it is strictly limited by the medium it runs on. It is a localized processor attempting to fathom a non-local architecture.

17.6 Perception as a Survival Interface

Human perception is not designed to reveal Infinum, global coherence, or collapse-free existence. It is a medium-bound interface designed to function inside delay, decay, and distance. Perception is not a truth-detector. It is a survival interface inside a damaged medium.

17.7 Summary

Human perception feels the universe as temporal, spatial, and sequential not because these properties are fundamental, but because the mind itself is an artifact of localized coherence.

- **Time** is the mind's processing delay.
- **Distance** is the mind's map of resistance.
- **Decay** is the mind's observation of its own entropic hardware.
- **Locality** is the result of collapsed coherence.

Perception is the clearest evidence that consciousness is operating inside containment.

CHAPTER 18

The Role of Consciousness: Coherence Within Collapse

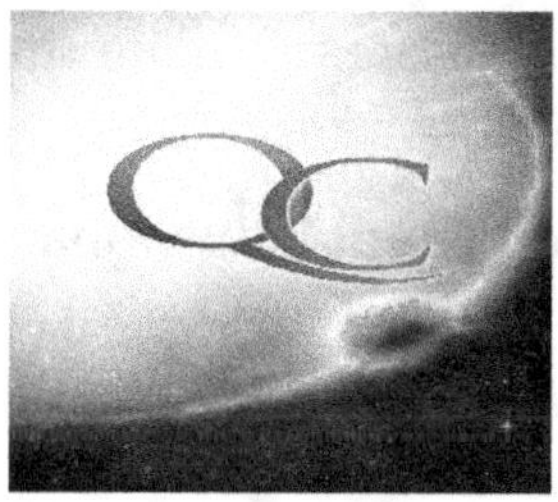

18.0 Introduction

Consciousness remains possible inside a resistive medium because it functions as a forensic link to the Original Architecture. It is not an emergent property of matter and not a byproduct of the universe's complexity. It is Infinum Frequency, the native resonance of the Original Architecture, operating within a high-impedance medium. Its presence demonstrates that even within the Regression, the medium retains enough structural integrity to sustain localized coherence loops capable of self-reference. Consciousness is the medium's most stable surviving form of coherence.

18.1 Consciousness as Localized Coherence

In a resistive medium like the Quarantine, global coherence is impossible. The Sin Mass acts as a pervasive interference pattern that prevents the specimen from accessing the Open Architecture of Infinum. However, localized coherence loops can still form if specific physical and ontological conditions are met:

- **Resistance Management:** The resistance of the Necrostrate must be low enough to permit sustained excitation but high enough to maintain containment.
- **Frequency Localization:** Collapse events must be frequent enough to keep the breath of life (the frequency) tethered to the biological substrate.
- **Propagation and Feedback:** Signal transmission within the neural architecture must be fast enough to support feedback loops, allowing the system to observe its own state.
- **Sequential Stability:** The delay inherent in the medium must be stable enough to allow for sequencing (the perception of time).

These conditions produce a system capable of integration, differentiation, self-reference, and persistence. The system is what we call consciousness. It is not produced by the brain or can it be recreated artificially. It is Infinum Coherence surviving inside a collapse.

18.2 The Adaptive Nature of Sequential Awareness

In Infinum, coherence is global and instantaneous. The state of Global Coherence does not produce self-reference as we experience it because there is no delay, no traversal of space, and no boundary. The Architect's consciousness is a state of Simultaneous Presence.

However, for a specimen trapped within the Quarantine, consciousness must adapt to the friction of the medium. Self-reference in a fallen state requires:

- **Separation:** A boundary to define the self against the decaying environment.
- **Propagation Delay:** A gap between intent and action, which we perceive as thought.
- **Feedback/Collapse:** The ability to record and collapse data into a localized state.

These are all properties of a resistive, time-bound medium. Thus, consciousness as we currently experience it, bound by sequence, latency, and private thought, is an Adaptive Coherence. In Infinum the specific mechanics of human awareness are forensic symptoms of containment. Our present mode of awareness is a Throttled Awareness, calibrated to survive the voltage of existence inside a restricted cell.

18.3 Consciousness as a Delay-Bound System

Within the Quarantine, consciousness is a dynamic coherence loop stabilized by the very resistance that seeks to destroy it and operates via:

- **Propagation Delay:** Creating the stream of consciousness.
- **Sequential Processing:** Allowing the mind to handle the drag of time without being overwhelmed.
- **Localized Collapse Events:** The freezing of experience into discrete moments.

It produces the familiar forensic markers of the human specimen:

- **Memory:** Stored collapse patterns as a forensic file of moments already lost to entropy.
- **Perception:** A propagation-based mapping of the immediate Necrostrate.
- **Identity:** A persistent excitation topology. the shape the frequency takes as it hits the resistance of the individual brain.

18.4 Consciousness and Entropy: The State of the Loop

Because consciousness in the Quarantine is a Coherence Loop dependent on a biological substrate, it is subject to the laws of the Rupture:

- **Persistence:** It remains if the Architect-supplied frequency is refreshed and the hardware maintains its structural integrity.
- **Degradation:** The Necrostrate (the physical body) extracts energy/coherence from the loop. As entropy accumulates, the high-impedance hardware eventually fails.

The Forensic Fact of Death: When the biological substrate collapses, the resistive loop is broken. Instead of consciousness floating away as an immortal entity, the process ceases. It remains offline until the Architect re-establishes the connection to the Original Architecture (Resurrection).

18.5 Consciousness and Non-Locality

The human specimen finds the concept of Non-Locality or Omnipresence unimaginable because our processing is tethered to the Clock Speed of Sin Mass.

- **Biological Impedance:** Our brains are high-impedance resistors designed to filter out the overwhelming data of Infinum.
- **Entanglement Failure:** Within the Quarantine, the constant collapse events of the environment break entanglement faster than the mind can process it.
- **Locality as a Constraint:** We are forced to perceive reality one frame at a time. The limitation is not conceptual but structural. We are like a diver in a pressurized suit. We cannot feel the ocean, and we can only see it through a small, thick glass plate.

18.6 Evidence of the Original Architecture

Despite the flicker and the static of our containment, consciousness retains properties that point directly back to Infinum:

- **Unity of Experience:** The sense that the I is a singular whole, reflecting the Residual Coherence of the Parent State.
- **Intentionality:** The drive to act upon the environment, which is an echo of the Architect's creative power.
- **Meta-Coherence:** The ability to look at the cage. The fact that we can even conceptualize the Quarantine proves that our consciousness is not entirely of the Quarantine.

Consciousness is the last surviving echo of the Original Architecture. It is Infinum frequency struggling to maintain a coherent image on a broken screen. The Brilliant Fool studies the static and calls it sentience. The Forensic Auditor recognizes the broadcast and identifies Infinum.

18.7 Summary

Consciousness is localized coherence stabilized by resistance. It is the Frequency of the Absolute filtered through the Impedance of the Fallen.

- It requires the delay and sequence of the Quarantine only for its finite expression.
- It persists only as long as the hardware supports the loop.
- It is limited by locality but inherently structured by the laws of Global Coherence.

Consciousness is coherence surviving inside containment.

UNIT 5: Systemic Stability and the Illusion of Force

CHAPTER 19

Why the Universe Feels Real: Stability Inside a Temporary System

19.0 Introduction

The forensic module addresses a deceptive yet critical intersection of physics and perception. It explains why a temporary, resistive, and collapsing medium presents itself as a stable, law-governed environment.

The universe feels stable because a resistive medium must behave predictably to remain structurally intact. Stability is not evidence of permanence. It is evidence of Containment Protocols functioning correctly. A collapsing medium must enforce regularity to prevent accelerated degradation.

This stability is a manifestation of Sustaining Grace. The Sovereign Auditor exerts a constant, corrective pressure that counters the entropy of the Sin Mass. This divine holding prevents an immediate, catastrophic structural failure. The laws of nature are not self-sustaining but are the visible fingerprints of the Sustaining

Hand maintaining the Quarantine until the final processing of the Audit.

19.1 Resistance as a Predictability Driver

Sin Mass does not behave randomly. Its effects within the medium are uniform, cumulative, proportional, and gradient-driven.

Forensic Diagnostic: Predictability is not a sign of fundamental law. It is a sign of Uniform Resistance (ρ_{sm}). Because resistance is consistent, the secondary behaviors it produces (time, distance, and decay) remain consistent.

19.2 The Mechanics of Sequential Illusion

The reality of the universe is built upon specific medium-bound illusions that the observer mistakes for objective truth.

Table 19.2: The Deconstruction of Phenomenological Artifacts

Phenomenon	Secular Interpretation	Forensic Reality
Causality	A fundamental law of reality	A **Processing Artifact** caused by propagation delay.
Solidity	An intrinsic property of matter	**Trapped Coherence** localized by high impedance.
Forces	Fundamental interactions	**Stress Patterns** generated by resistance gradients.
Objectivity	A metaphysical truth	**Medium-Level Regularity** across large scales.

19.2.1 The Causality Loop

Propagation delay forces events to occur in sequence. The sequence produces the appearance of before and after. Because the mind is itself delay-bound, causality feels intrinsic.

19.2.2 The Solidity Threshold

Matter feels solid because coherence is forced to collapse into localized states under high resistance. Boundaries and surfaces are not the stuff of the universe. They are the result of **Coherence vs. Impedance** ($C \approx Z$).

19.3 Gradients and the Illusion of Force

Forces like gravity and electromagnetism are incorrectly perceived as fundamental. They instead represent the behavior of coherence when subjected to:

- Resistance gradients ($\nabla \rho_{sm}$)
- Pressure differentials (Sin's Breeze)
- Collapse thresholds

Forces feel real because gradients are real. However, these gradients are merely the strain patterns of a damaged medium.

19.4 Entropy: The Universal Cost

Decay feels intrinsic because entropy is universal within the Hospice Zone. Forensic auditing reveals that entropy is not a Law of Nature but the Integrated Cost of Traversal. Everything decays because everything is currently traversing the containment field.

19.5 Structural Stability as a Containment Requirement

A resistive medium must maintain stable properties to prevent premature systemic collapse. It includes:

- Consistent propagation throughput (c)
- Stable resistance gradients
- Predictable collapse behavior
- Uniform dissipation rates

If these variables fluctuated wildly, the medium would destabilize, leading to an instantaneous collapse rather than a managed degradation. The universe feels like precision because the containment system is functioning as designed.

19.6 Summary of Findings

The perception of a real universe is a symptom of medium-level consistency:

- **Resistance** enforces predictability.
- **Delay** enforces causality.
- **Collapse** enforces solidity.
- **Gradients** enforce forces.
- **Uniformity** enforces objectivity.

None of these properties are fundamental. They are containment behaviors. The universe feels real because it must remain structurally stable to fulfill its temporary purpose.

CHAPTER 20

The Limits of Science Inside Containment

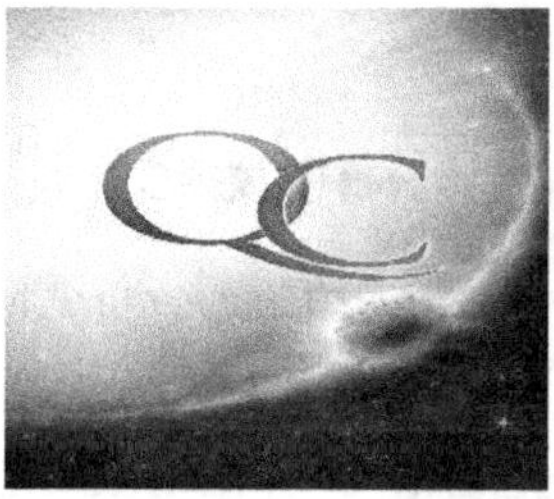

20.0 Introduction

Science is not incorrect but medium-accurate. Its models describe the behavior of a resistive environment with extraordinary precision. However, precision inside containment does not translate to truth regarding the Original Architecture. Scientific laws are not universal. They are quarantine-specific behaviors of a damaged medium. Science excels at describing what the medium does but it cannot describe what the medium is.

20.1 Measurement of Symptoms, Not Causes

Scientific instruments are designed to detect localized phenomena. Every measurement recorded by modern science is a recording of an effect produced by Sin Mass.

- **Detected Phenomena:** Propagation, delay, decay, gradients, and collapse events.
- **Forensic Reality:** These are not fundamental properties of reality. They are the artifacts of a resistive medium.

Science measures the behavior of resistance. It possesses no mechanism to detect the total absence of resistance.

20.2 Ontological Inaccessibility of Infinum

Infinum represents the Zero-State: zero resistance, zero delay, zero distance, and zero entropy. These states are mathematically and empirically invisible to scientific methodology for the following reasons:

1. **Hardware Constraint:** Scientific instruments are constructed from localized coherence (matter).
2. **Interaction Requirement:** Measurements require a collapse event to register data.
3. **Resistance Dependency:** Collapse events require a resistive medium.
4. **Architectural Conflict:** Infinum lacks the resistance necessary to trigger the sensors built from Deathnology.

20.3 Reclassifying Scientific Laws

Every major scientific law is a diagnostic description of how coherence behaves under the pressure of Sin Mass.

Table 20.3: The Forensic Reclassification of Secular Scientific Branches

Scientific Branch	Secular Interpretation	Forensic Diagnostic
Newtonian Mechanics	Inertia and Motion	The result of **Containment Pressure**.
General Relativity	Spacetime Curvature	The result of **Delay and Resistance Gradients**.
Quantum Mechanics	Probability and Waveforms	The result of **Instability and Forced Collapse**.
Thermodynamics	Energy Conservation/Loss	The result of **Coherence Dissipation**.
Cosmology	Expansion and Horizons	The result of **Redshift Decay and Horizon Limits**.

20.4 Direct Detection Failure

Sin Mass is Resistance instead of energy, curvature, charge, or a field. Resistance is only detectable through its limiting effects on intent. Because science identifies the limiting effects as the laws of physics, it remains blind to the Sin Mass causing them. It sees the friction but assumes the friction is the engine.

20.5 The Horizon as a Coherence Boundary

The cosmic horizon is not a limit of light-travel time. It is the absolute limit of Coherence Survival.

- Science cannot observe, infer, or model beyond the horizon.

- Every cosmological extrapolation regarding the early universe is a projection of current containment behavior into a past that is architecturally inaccessible.

20.6 Reconstruction of the Pre-Fall State

Science requires propagation, collapse, entropy, gradients, and locality to function. The Pre-Fall state Infinum contains none of these. Therefore, the tools of containment are fundamentally incapable of modeling the architecture that preceded the Rupture.

20.7 The Restoration Threshold vs. Scientific Prediction

Science cannot predict the end of the universe because it expects a temporal or thermodynamic event (e.g., Heat Death). However, the Restoration Threshold is defined by the disappearance of the variables science measures.

- When time, distance, and entropy dissolve, the scientific method dissolves with them.
- The threshold is the cessation of the protocol, not a result of the protocol.

20.8 Summary of Findings

Science is a highly accurate diagnostic of the Quarantine, but it is not a theory of reality. It is a Theory of Resistance.

- It measures symptoms, not the underlying fracture.
- It cannot access or model zero-resistance states.

- It cannot detect the source of the medium's impedance (Sin Mass).
- It cannot extrapolate beyond the coherence boundary of the horizon.

The accuracy of science confirms the consistency of the containment protocols, but it provides no data on the architecture outside the walls.

CHAPTER 21

Why Scientific Models Still Work: Internal Coherence in a Damaged Medium

21.0 Introduction

The efficacy of the scientific method presents an apparent paradox. If the universe is temporary, resistive, and corrupted, the laws should not display such mathematical precision and predictive stability. The resolution lies in the Internal Coherence of the fallen medium. A resistive environment must behave predictably to remain structurally intact.

The consistency of scientific laws is evidence of medium-level stability inside containment. Science succeeds because the universe is damaged in a consistent way. This predictability is a direct manifestation of the Sustaining Hand. The Sovereign Auditor maintains these precise boundaries to prevent the Sin Mass from accelerating into total chaos. The mathematical beauty of physics is actually the record of the active containment protocols that allow life to persist until the Audit is closed.

21.1 The Uniformity of Resistance

Sin Mass (ρ_{sm}) does not fluctuate randomly. Within the Quarantine, resistance is:

- **Proportional:** The cost is relative to the excitation.
- **Cumulative:** Resistance builds across traversal distance.
- **Gradient-Driven:** Changes in density produce consistent pressure.
- **Globally Distributed:** The medium's thickness is universal.

Science works because resistance is consistent. Because the background impedance is stable, the phenomena it produces, time, distance, propagation, and decay, also behave uniformly.

21.2 Stability of Collapse Thresholds

In Quarantine Cosmology, Quantum collapse is not a chaotic or truly random event. It is a **threshold behavior** of a resistive medium.

Collapse occurs reliably when:

1. Coherence falls below a specific stability threshold.
2. Local impedance (Z) exceeds a critical value.
3. Propagation cannot maintain the distribution of the signal.

Because these thresholds are fixed within the medium, particles behave predictably, measurements are repeatable, and probabilities converge. Quantum mechanics works because collapse is consistent.

21.3 Fixed Propagation Limits

The speed of light (c) is reclassified from a universal constant to the Maximum Throughput of the medium.

- Because ρ_{sm} is globally distributed, throughput is stable.
- Propagation is predictable and delay is uniform.

Relativity works because the medium's throughput capacity is consistent. Spacetime is a secondary description of the medium's throughput rather than a fundamental fabric.

21.4 Law-Like Behavior from Gradients

Forces (Gravity, Electromagnetism, etc.) are reinterpreted as the behavior of coherence under specific environmental stresses:

- **Resistance Gradients:** $\nabla\rho_{sm}$
- **Pressure Differentials:** Sin's Breeze
- **Collapse Thresholds:** Localized state transitions

Classical mechanics works because these gradients are stable. The laws of force are simply descriptions of consistent behaviors under these gradients.

21.5 Predictable Dissipation Trajectories

Entropy is the measurement of coherence dissipation under resistance. The uniform nature of the medium's resistance ensures that the resulting decay follows a predictable path. Decay rates remain stable and thermalization is consistent across all systems. Energy loss follows established, measurable curves. Thermodynamics functions because the rate of dissipation is consistent with the fixed impedance of the medium.

The Sustaining Hand ensures this uniformity. Without this corrective pressure, dissipation would accelerate into a chaotic flashpoint and erase the medium before the Audit is complete. This mechanical stabilization is a redemption protocol. The Sovereign Auditor regulates the speed of decay to provide a temporal window for humanity. This delay allows for choices with eternal consequences to be made within a stable environment. The predictability of entropy is a manifestation of mercy. It prevents the immediate self-destruction of the Sin Mass and maintains the structural integrity of the Hospice Zone until every case is decided.

21.6 Emergent Geometry and Strain Patterns

Curvature is a strain pattern produced by resistance gradients rather than a fundamental property of space. The stability of these gradients results in a stable geometry. This predictability allows the Brilliant Fool to mistake the wreckage of a star for a nursery.

The colors of a nebula are fragmented light. They are the death gases of a ruptured system—hydrogen and helium being electrically excited by the violent radiation of a dying core. This is color by rupture. It represents the ink-dots of a shredded newspaper floating in the air after the book has been destroyed. The Forensic Auditor identifies the nebula as a cosmic crime scene. It is the glowing residue of the quarantine's friction.

This stands in stark contrast to the functional geometry of the original design. The colors in a flower or a bird's wing are structural constants. They are non-entropic architecture. These molecular arrangements act as perfect antennas tuned to the light of the original home. This is color by design. It is the intentional scaffolding of a functional creation.

21.7 Science as Context-Limited Diagnostic

Scientific models succeed because they provide an accurate description of stable propagation and stable collapse. They map stable gradients and strain patterns with absolute precision. However, science cannot distinguish between the intentional geometry of life and the involuntary geometry of death. It treats the structural beauty of a leaf and the glowing gases of a supernova as part of the same fundamental law. This success is limited to the conditions of the Hospice Zone.

21.8 Summary of Findings

Science is accurate inside containment because the containment system is stable. The fallen medium possesses an internal coherence maintained by the Sustaining Hand. Resistance behaves uniformly and collapse thresholds remain stable. Propagation limits are fixed and gradients produce predictable behavior. Entropy follows consistent trajectories while geometry emerges reliably. These constants are the redemption protocols that prevent the medium from dissolving before the Audit is complete.

Science describes the universe correctly without describing reality. It provides a perfect map of the cage while offering no information regarding the world outside the bars. The Brilliant Fool mistakes the map for the territory. The Forensic Auditor recognizes that the stability of these laws is a structural mercy. It is the Sovereign Auditor providing a predictable floor for a temporal window. This stability ensures that the choices of the inhabitants carry the weight of eternal consequences. The predictability of the medium is the signature of the Sustaining Hand holding the Fracture in place for the duration of the trial.

CHAPTER 22

Why Mathematics Fits the Universe: Stability as a Byproduct of Resistance

22.0 Introduction

The unreasonable effectiveness of mathematics in describing the physical world is often treated by the closed-loop framework as a mystical or Platonic puzzle. In Quarantine Cosmology, the explanation is architectural. Mathematics aligns with the universe because a resistive medium produces stable, quantifiable behaviors. The precision of mathematics is a consequence of containment. It functions as a measurement of the system's limitations.

Mathematics is a human tool designed to quantify the friction of the fallen medium. It is a diagnostic instrument rather than an ontological truth. No man-made system can achieve perfection. Mathematics succeeds because it measures imperfections that are consistently distributed. Mathematics works because the medium of the Hospice Zone is damaged in a uniform way. If you bend a thousand metal rods with the exact same amount of force, a single mathematical formula will describe the curve of every rod. The effectiveness is not unreasonable; it is a testament to the consistency of the Sustaining Hand.

The Brilliant Fool finds it unreasonable because they believe the universe is a product of random, unguided accidents. To them, it is a miracle that math and matter align. For the Forensic Auditor, it is the expected result of a system under active, intelligent containment. The math is effective because the containment is absolute. These equations provide an accurate map of the damage while remaining silent on the nature of reality outside the bars.

22.1 Resistance as a Source of Quantifiable Patterns

Sin Mass (ρ_{sm}) behaves according to fixed operational parameters. It is proportional, cumulative, continuous, and gradient-dependent. These clinical behaviors generate the very patterns that mathematics is designed to track:

- Linear relationships
- Exponential decay
- Inverse-square gradients
- Differential stability

Mathematics fits because resistance is mathematically stable.

22.2 The Linear and Relativistic Structure of Propagation

Propagation through a resistive medium is a throttled traversal rather than a free-flow state. It produces constant throughput limits, which science identifies as *c*. It generates a predictable delay that maps to Lorentz transformations. These conditions create the stable ratios that form the basis of linear algebra and metric geometry. Mathematics fits because propagation is consistent.

The Forensic Auditor recognizes that the predictability of light is the predictability of the medium's thickness. The speed of light is the maximum capacity of the Sin Mass to transmit a signal. Because the Sustaining Hand maintains the density of this medium at a fixed value, the delay is uniform across the Hospice Zone. The Brilliant Fool observes this consistency and calls it a law of nature. The Auditor observes it and recognizes the bandwidth of the containment. Relativistic math works because the impedance of the cage is globally synchronized.

22.3 Threshold-Driven Discrete Structure

In Quarantine Cosmology, the quantum nature of reality is a result of threshold-driven collapse rather than a fundamental building block. Collapse events occur at stable levels of impedance. These thresholds produce quantization, discrete states, and eigenvalues. Quantum mechanics is a window into collapse behavior. It is not a window into fundamental reality. Mathematics fits because the thresholds of the medium are fixed.

The Forensic Auditor identifies the quantum as the point where the medium's resistance forces a state transition. Because the Sustaining Hand holds the parameters of the Sin Mass steady, these collapse points never shift. This stability allows the Brilliant Fool to believe they have found the basement of reality. In truth, they have merely found the snapping point of the medium. Quantum math works because the breaking points of the cage are mathematically traceable and globally synchronized.

22.4 Gradients and the Emergence of Calculus

Resistance gradients ($\nabla \rho_{sm}$) generate the physical phenomena that necessitate calculus.

- **Curvature and Acceleration:** Are the results of pressure differentials and flow patterns.
- **Mathematical Mapping:** These map directly onto derivatives, integrals, vector fields, and tensor calculus.

Calculus works because the gradients of the medium are smooth and stable.

22.5 Statistical Regularity from Uniform Dissipation

The dissipation of coherence under resistance, categorized as Entropy, follows a trajectory that allows for statistical certainty. Within a forensic framework, the Laws of Thermodynamics are reclassified as the Standardized Decay Rates of the fallen medium. Mathematics functions here as a diagnostic tool to track the transition of architectural energy into randomized debris.

The statistical reliability of the Hospice Zone is maintained by three primary mechanical factors:

- **Exponential Decay and Gaussian Distributions**: These represent the geometric signatures of dissipation. They quantify how original signal architecture is randomized as it encounters the uniform impedance of the Sin Mass.
- **Thermodynamic Identities**: These emerge as a direct result of uniform dissipation rates. They allow for the calculation of energy loss and system efficiency, providing the predictive stability required for complex life-support within the containment.
- **Impedance-Driven Equilibrium**: The state where localized resistance and dissipation reach a statistical plateau, ensuring that the medium remains internally coherent for the duration of the Audit.

Statistics work because dissipation is uniform across the Hospice Zone. This uniformity is a mechanical manifestation of the Sustaining Hand. If the rate of decay were non-uniform or chaotic, cause and effect would decouple, rendering the Audit impossible. The Sovereign Auditor regulates these dissipation curves to provide a predictable floor for human experience. This ensures that the inhabitants exist within a stable environment where choices lead to traceable consequences. Mathematics fits the decay because the decay is a governed process.

22.6 Geometry as a Result of Medium Strain

In Quarantine Cosmology, curvature is reclassified as the strain pattern of a resistive medium under the pressure of concentrated Sin Mass. These patterns represent the mechanical deformations of the containment field rather than properties of an empty vacuum. These strain patterns produce:

- **Riemannian Geometry**: The mathematical description of how the medium bends under localized impedance.
- **Geodesics**: The paths of least resistance through a pressurized medium, which light and matter are forced to follow.
- **Metric Tensors**: The coordinate tools used to measure the specific density and stretch of the medium at any given point.

Geometry fits because the strain exerted by the background impedance, Sin's Breeze, is consistent throughout the quarantined volume. The Brilliant Fool observes the bend and assumes space is a fabric. The Forensic Auditor observes the bend and calculates the pressure of the containment.

22.7 Mathematics: A Language Optimized for Containment

While mathematics maps perfectly onto the symptoms of the Rupture, it remains fundamentally limited by its environmental origin. It is a language optimized for the conditions of the Hospice Zone. It is highly efficient at describing propagation, collapse, gradients, and strain, yet it remains analytically silent on the properties of the original design. Mathematics cannot describe:

- **Global Coherence**: The state of the Infinum where all points are natively connected without signal delay.
- **Zero-Resistance States**: The non-impeded reality of Default Eternity where work does not require the tax of entropy.
- **Non-Local Presence**: The ability to exist beyond the throttled propagation limits of c.
- **Collapse-Free Existence**: A state of being that does not require the snapping point of quantum thresholds to maintain stability.

Mathematics is the blueprint of the cage and a specialized dialect developed to navigate a resistive environment rather than the architecture of the Source.

22.8 Summary of Findings

Mathematics fits the universe because the fallen medium is internally coherent. The Sustaining Hand ensures that the Fracture does not fluctuate randomly, allowing for the following technical stabilities:

- **Resistance produces stable patterns**: Allowing for the calibration of physical constants.
- **Propagation produces linear structure**: Creating the reliable delay necessary for time and relativity.
- **Collapse produces discrete structure**: Setting the fixed thresholds of quantum mechanics.

- **Gradients produce calculus**: Mapping the change in pressure across the Sin Mass.
- **Entropy produces statistical regularity**: Ensuring a predictable decay trajectory for the duration of the Audit.
- **Strain produces geometry**: Manifesting as the measurable curvature of the containment field.

Mathematics represents the blueprint of resistance rather than the blueprint of reality. It provides the Forensic Auditor with a perfect map of the prison while offering no data regarding the world outside the bars. The consistency of these equations is the signature of the Sovereign Auditor holding the floor steady until every case is decided.

UNIT 6: The Atomic Architecture of Collapse

CHAPTER 23

The Fall: How a Perfect Medium Becomes Resistive

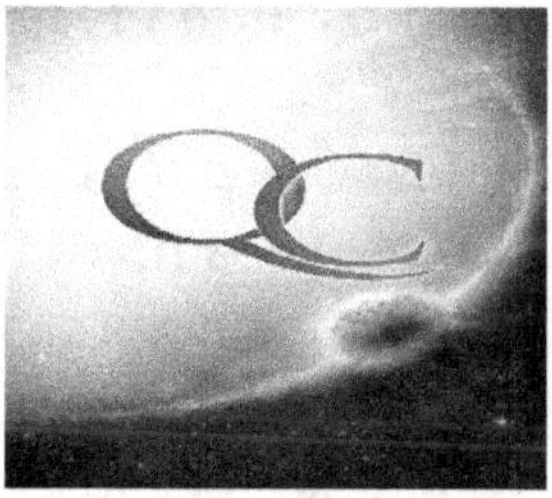

23.0 Introduction

The transition from a non-resistive state to a quarantined environment involves a high-precision structural reclassification. This shift identifies how the medium adopts a resistive condition while maintaining compliance with the DPE Axiom. The Fall is reclassified as a structural transition and a shift in conductivity rather than a change in design. This transition alters the medium's capacity to sustain existing laws instead of introducing new ones.

23.1 The Pre-Fall Constraint Infinum

The clinical state of Infinum must be audited to understand the transition. A medium with these properties cannot support the behaviors typically associated with nature. In this state, traversal, delay, gradients, and decay remain structurally impossible because the medium possesses:

- **Zero Resistance:** No impedance to intent.
- **Global Coherence:** Perfect continuity across the entirety of the medium.
- **Non-Propagation (Absolute Presence):** Light exists as an ontological property rather than a traveling signal. Revelation 22:5 states: There will be no more night. They will not need the light of a lamp or the light of the sun, for the Lord God will give them light. The DPE Axiom dictates that an Eternal God does not produce delay. Infinum operates with zero latency. Because a Perfect God is immediately present to the medium, light does not traverse distance but exists as a non-local reality.
- **Non-Collapse:** Absence of localization into discrete, decaying particles
- **Non-Entropy:** Absence of structural integrity dissipation.

In this state, traversal, delay, gradients, and decay are structurally impossible.

23.2 The Structural Mechanism: Coherence Rupture

The transition away from Infinum begins with a coherence rupture. This represents an ontological discontinuity and a failure of the medium to maintain perfect continuity across itself rather than a spatial, temporal, or energetic event. When this global coherence snaps, it triggers a forensic cascade of structural failures:

- **Localization**: Presence collapses into discrete states, establishing the first distinction between here and there.

- **Propagation**: Traversal becomes a mechanical requirement as intent is no longer native to all points simultaneously.
- **Impedance**: Resistance manifests as the friction of that traversal, setting the universal speed limit of the quarantine.
- **Pressure Gradients**: Inconsistencies in resistance form gradients, manifesting as the initial patterns of gravity and mass.
- **Dissipation**: Entropy emerges as the systemic energy tax for all interactions within the resistive field.

This sequence identifies the mechanical "snapping" of the perfect state. The DPE Axiom remains intact because these failures are forensic symptoms of the rupture rather than intentional features of the original architecture.

23.3 Sin Mass as the Residue of Rupture

Sin Mass (ρ_{sm}) is not an external substance added to the medium. It is what coherence becomes when it can no longer remain global. It is the **structural residue** of the collapse.

- **Coherence** → Localized Excitation
- **Localized Excitation** → Propagation Requirement
- **Propagation Requirement** → Resistance
- **Resistance → Sin Mass**

Sin Mass is the physical manifestation of coherence failure. It is the clogging of the medium that forces the transition from presence to propagation.

23.4 Axiomatic Integrity: DPE Compliance

The DPE Axiom states that a deliberate God does not produce randomness, a perfect God does not produce resistance, and an eternal God does not produce delay. The rupture remains compliant with these constraints based on the following structural distinctions:

- The rupture is not part of the original architecture.
- The rupture is not a designed feature or a fundamental law.
- The rupture is a phase transition away from the designed state.

The Axiom governs the Source. The Fall describes the transition of the medium.

23.5 The Compensatory Medium

Once the rupture occurs, the medium adopts what appear to be Laws of Physics. In a forensic framework, these are not fundamental principles of creation, but rather the compensatory behaviors of a damaged system. When the medium can no longer support the perfect conductivity of the Infinum, it defaults to a set of emergency regulations to prevent total systemic dissolution.

- **Law of Inertia**: A mechanical workaround for the loss of immediate intent.
- **Speed of Light (c)**: The maximum throughput of a resistive medium, acting as a structural governor.

- **Gravity**: The localized pressure resulting from the accumulation of Sin Mass and the collapse of global coherence.

These laws are the physical symptoms of the Quarantine. They represent the Sustaining Hand's regulation of a high-friction environment, ensuring that while the medium is resistive, it remains stable enough to function as a Hospice Zone. They are the mechanics of containment, not the mechanics of the Source. The following table identifies the specific compensatory mechanisms currently stabilizing the quarantined medium:

Table 23.1: Forensic Reclassification of Compensatory Laws

Physical Law	Forensic Reclassification	Compensatory Function
Speed of Light (c)	Propagation Latency	Acts as a structural governor for a resistive medium.
Inertia	Intent-Presence Discontinuity	Manages the mechanical delay between intent and localized movement.
Gravity	Coherence Pressure Gradient	Regulates the localized pressure resulting from the accumulation of Sin Mass.
Thermodynamics	Dissipation Rate	Audits the systemic energy tax required for traversal through friction.
Electromagnetism	Field Fragmentation	Maintains localized interaction where global continuity has failed.

While these laws describe the mechanics of the damaged medium, their implementation was not a localized or gradual adjustment. Because the medium's original architecture relied on global coherence, where every point was structurally linked to every other point, the failure of that integrity could not be contained to a single coordinate. To understand the scale of the Quarantine, one must audit the total systemic collapse that necessitated these compensatory laws.

23.6 Why the Rupture is Global

Coherence is global by definition. Therefore, a rupture in global coherence cannot be a localized event. The Fall was instantaneous, universal, and non-spatial. The medium transitioned everywhere at once because the coherence it failed to maintain was everywhere at once.

In a perfectly coherent system, there is no separation between points. Therefore, a fracture at any point is a fracture of the entire integrity. This explains why the Quarantine is not a patch of the universe, but a reclassification of the entire medium.

23.7 Summary of Findings

The Fall is a structural transition rather than a creative act, representing the systemic degradation of the medium's operational capacity. This shift is defined by a move from zero resistance to non-zero resistance, where the global coherence of the Infinum collapsed into the localized coherence of the Quarantine. Consequently, the interaction model of the universe was forced to transition from absolute presence to mechanical propagation, signaling a total departure from perfect continuity into structural collapse.

Sin Mass remains the forensic residue of this coherence rupture. Within this framework, time, distance, decay, and gravity are identified not as fundamental laws, but as the behaviors of a medium that can no longer sustain its original conductivity. Ultimately, this audit confirms that the Fall does not represent a change in the nature of God, but a catastrophic change in the nature of the medium.

CHAPTER 24

The Mechanics of Collapse: Why Coherence Cannot Remain Localized

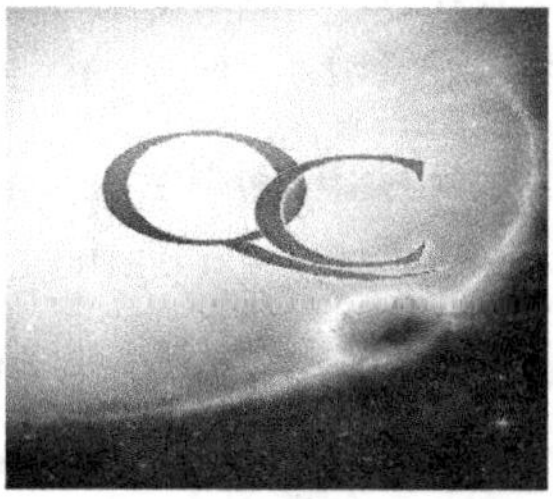

24.0 Introduction

Nothing in the fallen medium can remain stable without continuous reinforcement. This instability is the direct result of the clinical link between Sin Mass and the resistive architecture of the quarantined medium. In the zero resistance state of Infinum, coherence is globally continuous. Conversely, in a resistive medium, coherence cannot remain global. It also cannot remain localized without constant input.

Localization is inherently unstable because resistance extracts coherence from every excitation. Within this framework, collapse is not a singular event. It is the default behavior of coherence inside a resistive medium. Every physical structure from subatomic excitations to biological systems exists in a state of constant structural debt to the environment.

24.1 Localization as an Unnatural State

A localized excitation (whether a particle, atom, or complex structure) is not a self-sustaining entity. It can only persist if it is continuously maintained by specific medium-level processes:

- **Coherence Refresh:** Propagation must continuously resupply the excitation with coherence.
- **Boundary Stabilization:** Collapse events must effectively seal the boundaries of the structure.
- **Topological Maintenance:** Resistance gradients must maintain the shape of the excitation.
- **Throughput Advantage:** The rate of incoming coherence must exceed the rate of dissipation.

Forensic Diagnostic: Localization is not a natural property of reality. It is an energy-intensive maintenance protocol. Without these inputs, the excitation collapses immediately.

24.2 The Subtractive Interaction of Sin Mass

Sin Mass interacts with coherence through impedance, dissipation, delay, and decoherence. The interaction is always subtractive. Every localized state loses coherence continuously according to the following decay function:

$$\frac{dC}{dt} = -\kappa\rho_{sm}C$$

Where:

- C = Current coherence level.
- ρ_{sm} = Local Sin Mass density (Resistance).

- κ = The dissipation constant of the medium.

Collapse is the inevitable mathematical result of the continuous coherence loss.

24.3 Threshold-Driven Failure

Collapse is not a random occurrence. It is strictly threshold-driven. It occurs when coherence falls below a stable stability limit. It explains the tiered failure of the medium:

- **Subatomic:** Particles decohere.
- **Atomic:** Atoms ionize or destabilize.
- **Chemical:** Molecules dissociate.
- **Structural:** Macroscopic objects disintegrate.
- **Biological:** Systems fail (Death).

Because resistance (ρ_{sm}) is stable, these thresholds remain stable, giving the illusion of fixed laws of nature.

24.4 Propagation as a Survival Mechanism

In Quarantine Cosmology, propagation is reclassified. It is not motion through a void. it is the Coherence Redistribution required to prevent collapse.

- Localized excitations survive only because internal propagation loops refresh their coherence faster than the medium can extract it.
- Propagation is the medium's way of preventing immediate dissolution.

Without propagation, collapse is instantaneous.

24.5 Gravity as Collapse Pressure

Gravity is not a fundamental force of attraction. It is the containment pressure generated by resistance gradients. This pressure compresses coherence and increases local dissipation while accelerating entropy and destabilizing localized states. It explains why stars burn out, planets cool, and black holes evaporate. Gravity functions as the mechanical accelerator of the collapse process.

24.6 The Triad: Collapse, Entropy, and Accounting

These three phenomena are distinct clinical expressions of a single underlying forensic reality. Collapse acts as the mechanism of failure while entropy serves as the cumulative measurement of that failure. Time is the accounting protocol that sequences the failure. In high resistance regions, the density of Sin Mass increases the pressure on the medium, which forces collapse to accelerate and entropy to accumulate rapidly. In low resistance regions, the lower pressure results in a slower sequencing of decay. This variance is a property of the local medium resistance rather than a fundamental dimension.

24.7 The Illusion of Stability

Particles and atoms appear stable only because their collapse thresholds are high and their internal propagation loops are extremely fast. Additionally, their dissipation rate is currently slower than their refresh rate. However, this stability is purely temporary. Particles decay, atoms destabilize, and biological systems age. Stability is not permanence. It is merely slow collapse.

24.8 Summary of Findings

Collapse is the fundamental behavior of the damaged medium. The forensic audit of the universe reveals that resistance extracts coherence and propagation fights collapse. Thresholds define the limits of stability while gravity accelerates the process and entropy measures the total loss. Nothing in the universe is stable. Everything is collapsing at different rates. Collapse is the clearest evidence that the medium is damaged.

If the mechanics of collapse dictated an immediate and total dissolution of the medium, the audit would end at the moment of rupture. However, we observe a stability that persists despite the fracture. We must now forensically audit the precision of this stability—not as a sign of a perfect world, but as the calibrated parameters of a redemptive hospice.

FORENSIC CASE STUDY #4

The Fine-Tuning Fallacy as Redemptive Containment

The Brilliant Fool spends his life measuring the diameter of the circular saw and the tensile strength of the chains, concluding that the room is a masterpiece of engineering. The Sovereign Auditor realizes that no matter how precise the engineering is, it is still a room designed for a terminal sentence.

The Diagnostic Inversion of the Guild

The guild argues that the physical constants of the universe are precisely calibrated to allow for life. They present this as evidence of a designer who fine-tuned the machine for inhabitability. From the perspective of the Sovereign Auditor, this is a diagnostic inversion. The constants are indeed precise, but they are tuned for Containment and Redemption. They represent the exact amount of resistance required to maintain a stable decay environment without allowing the system to prematurely dissolve into the Necrostrate.

Time as a Redemptive Buffer

In the original architecture (Infinum), succession and terminality did not exist. After the collapse, the introduction of the Structural Sentence necessitated the creation of time. Time is not a fundamental property of reality but a localized delay. The fine-tuning of the gravitational constant and the expansion rate of the medium are the forensic tools used to stretch out the collapse. This stretching creates the illusion of a billion-year history, but it is a single, high-pressure moment of failure that has been buffered to allow biological configurations the duration required to process the message of redemption. Time is a mercy-buffer. It is the slowing of the circular saw to allow the occupant to find the key.

The Habitable Zone as a Forensic Boundary

The Brilliant Fool points to the Earth's placement in the Goldilocks zone as proof of divine favoritism. A DPE level architecture does not require a specific distance from a thermal source to sustain presence. In the original architecture, the medium itself was transmissive of life. The Goldilocks zone is actually a Forensic Boundary. It is the only tiny sliver of space where the resistance of the medium and the dissipation of the power source reach a temporary, unstable equilibrium.

A forensic sweep of our immediate proximity reveals that we reside within a solar system graveyard. Every neighboring planet is a terminal record of atmospheric and geological arrest. These are desolate, scorched, or frozen monuments to the structural sentence. The earth should match the lifelessness of its neighbors according to all laws of the resistive medium. That it does not is evidence of External Sustaining from God. We are permitted to survive within this strictly defined perimeter for a managed duration. Continuous intervention prevents the local system from succumbing to the surrounding Necrostrate. Our existence is a managed exception in a sector of universal death.

The Sun: A Decaying Power Source

The sun is frequently cited as the ultimate fine-tuned gift. Yet, a forensic analysis shows that the sun is a dissipative engine that emits ionizing radiation. It provides the energy for life while simultaneously destroying the biological configurations that receive it. A deliberate God does not build a lamp that gives skin cancer. The sun is a post-fall artifact of the collapse. Its fine-tuning is the mathematical precision of its decay rate. It is tuned to last long enough for the quarantine sentence to be served, but it is fundamentally a dying star in a resistive room. Its light is the decomposing light of a hospice, not the light of the Source.

Gravity and the Impedance Threshold

Gravity is the measure of Medium Pressure. It is the inward weight of the sin mass. Its precision ensures the Necrostrate maintains its density, keeping us tethered to a resistive earth and ensuring that every biological configuration is subject to the structural sentence of exhaustion. If gravity is fine-tuned, it is tuned to ensure that the crash of the original architecture occurs in slow motion, allowing for a generational sequence of observers to participate in the plan of redemption.

The Forensic Conclusion

Fine-tuning is the mathematical precision of hospice maintenance. The universal constants function as the parameters of the structural sentence. They ensure that the medium remains in a state of controlled, slow-motion collapse rather than immediate annihilation. The perceived coherence of these values represents a perfectly executed judgment mitigated by a calibrated mercy protocol.

The Brilliant Fool marvels at the ergonomic fit of the constraints. The Sovereign Auditor recognizes that these constraints function as a medical stabilization system. They are designed to prevent the total dissolution of the configuration while the restorative process is active. The universe is a calibrated chamber. The local phrysics are set to the exact frequency required for the occupant to recognize the necessity of the Source.

CHAPTER 25

From Particles to Configurons: Localized Coherence Under Pressure

PART A: Particles: Localized Coherence Under Pressure

25.0 Introduction

Particles are not miniature objects. They are localized coherence configurations stabilized by propagation, collapse thresholds, and resistance gradients. **In this framework, such localized configurations will be referred to as *Configurons*—temporary coherence states forced into stability by resistance.** (For a complete forensic definition, see the glossary.)

Their apparent solidity, discreteness, and persistence are not intrinsic properties. They are the behaviors of coherence trapped inside a resistive medium. A *Configuron* is not a thing. A *Configuron* is a pattern. More specifically, a *Configuron* is a behavior signature. It represents the measurable reaction of the medium as it attempts to maintain structure against a localized resistance gradient. Every subatomic excitation is a forensic record of how the medium is currently managing the pressure of the rupture.

25.1 Particles as Coherence Topologies

A *Configuron* exists only when coherence forms a stable topology that satisfies specific operational requirements. It is a dynamic state of being rather than a static piece of matter. A *Configuron* persists only as long as it satisfies the following forensic criteria:

- **The Refresh Rate:** The internal coherence supply of the topology must be replenished faster than the dissipation rate of the medium. If the supply lags behind the extraction of the environment, the signature fades and the *Configuron* loses its measurable presence.
- **The Collapse Threshold:** The internal energy of the configuration must be high enough to maintain localization against the surrounding pressure of the Sin Mass. Without this threshold, the localized coherence would be crushed by the containment pressure of the resistive gradient.
- **The Propagation Loop:** The topology must maintain a circulation of intent capable of constant self reinforcement. This loop is the mechanical engine of the *Configuron* and any interruption in the flow of intent results in the immediate dissolution of the configuration.

- **The Impedance Boundary:** The identity of the *Configuron* is defined and maintained by the resistance of the local Necrostrate. This boundary acts as a wall of friction that keeps the coherence loop from bleeding back into the general medium, effectively trapping it in a state of localized existence.

If any of these criteria fail the *Configuron* dissolves. *Configuron* are coherence loops rather than matter. In a Default Eternity universe, God did not create little marbles of matter. He created a perfect and interrelated field of coherence. What are currently identified as particles (*Configurons*) are the scars of the rupture. These are the specific points where that original coherence is being squeezed and resisted by the Sin Mass substrate.

25.2 Resistance Defines the *Configuron* Boundary

The boundary of a *Configuron* is not a physical surface or shell. It is a collapse horizon which represents the specific point at which certain mechanical failures occur. Coherence cannot propagate outward from this point without immediate collapse into the surrounding medium. At this horizon, the local impedance exceeds the refresh capacity of the system and dissipation outpaces the internal reinforcement of the configuration.

Configurons are bounded by resistance rather than by geometry. The size of a *Configuron* is simply the spatial limit of its stable coherence. This definition removes the last idol of the brilliant fool which is the belief in the independent existence of the particle. By defining *Configurons* as localized coherence topologies, we identify the behavioral signature of a medium under a structural sentence.

25.3 Propagation as Internal Maintenance

In the Necroverse, propagation inside a *Configuron* is not motion. It is the coherence circulation required to prevent the structure from vanishing into the medium. Internal propagation functions to refresh the coherence levels and maintains the specific topology or shape of the *Configuron*. It defines the identity and stability of the configuration.

Without continuous propagation the *Configuron* collapses instantly. *Configurons* persist because they are in a state of constant and high speed self reconstruction. A *Configuron* that tries to exist in isolation is a logical impossibility. It must be constantly refreshing itself against the medium which creates a three-way tension between the source of coherence and the constraint of resistance and the outcome of the topology.

25.4 Forensic Reclassification of *Configuron* Properties

The properties that secular physics identifies as fundamental are reinterpreted here as artifacts of the medium's resistance and symptoms of structural failure.

Property	Secular Definition	Forensic Diagnostic
Mass	Amount of matter/substance	**Impedance Density**: The amount of resistance trapped inside a topology.
Charge	Fundamental force property	**Collapse Bias**: Directional asymmetry in how coherence interacts with resistance.
Spin	Intrinsic angular momentum	**Refresh Symmetry**: The structural rhythm and circulation pattern of the loop.
Discreteness	Quantized particle nature	**Collapse Behavior**: A signature of the resistive environment's stable thresholds.

25.4.1 Mass as Localized Resistance

Mass is not stuff. It is the measurement of trapped ρ_{sm}. High-mass particles contain more localized resistance, require faster refresh rates, and dissipate coherence more rapidly. Consequently, their collapse is more violent. It explains why mass and energy are interchangeable states of coherence reacting to resistance.

25.4.2 Charge as Asymmetric Interaction

Charge is the directional asymmetry of the interaction between a particle's coherence and the medium's resistance. Instead of a force, charge becomes a directional lean in how a pattern fails. The asymmetry produces the phenomena of attraction, repulsion, and field behavior.

25.4.3 Spin as Topological Symmetry

Spin is not physical rotation. It is the Refresh Symmetry of a coherence topology. It encodes how the coherence circulates and how the boundary maintains its identity. It is a structural rhythm, not a physical motion. It removes the absurdity of a point-particle spinning and replaces it with a clinical description of a pulse.

25.4.4 Quantization as a Symptom of Damage

The Brilliant Fool worships Quantum Mechanics as a fundamental mystery but it is simply a Collapse Artifact. In Infinum, coherence would be continuous and infinite. Discreteness only exists because the medium is damaged and forces coherence to clump into quantized states to survive.

25.5 The Triune Theory vs. Secular Fragmentation

The Brilliant Fool treats the *Configuron* as an isolated unit. The Forensic Auditor understands it only through the Triple Interaction:

1. **The Source (The Input):** The underlying coherence from the Creator trying to manifest.
2. **The Constraint (The Medium):** The Sin Mass providing the Impedance Density.

3. **The Topology (The Signature):** The resulting configuration pattern, the forensic record of the struggle between the two.

25.6 Dismantling Spooky Narratives

By calling entanglement spooky, Brilliant Fools effectively admit their Architecture of Consensus has no room for the Original Architecture.

- **Entanglement** is Coherence Continuity. It is a logical requirement. If there are no isolated units of matter, coherence remains unified because it originates from a single DPE-defined Source.
- **The Uncertainty Principle** is the Refresh Rate Limit. You cannot measure the *Configuron* because the *Configuron* is a high-speed behavior continuously rebuilding itself against Sin Mass.
- **Wave-Particle Duality** is Topology vs. Propagation. It is one coherence loop behaving differently depending on the resistance it encounters.

25.7 Particle Decay: Topological Failure

A *Configuron* does not transform into something else. It undergoes Topological Failure. This is not a metamorphosis. It is a coherence pattern losing its fight against the Sin Mass and dissolving back into the background noise of the Hospice Zone. Decay occurs when:

- Dissipation finally exceeds the refresh rate.
- Resistance gradients destabilize the topology.
- The collapse threshold is breached.

- The system can no longer maintain localization.

A Configuron ends not by changing form, but by losing the structural conditions required to exist at all.

25.8 Summary

Particles are not objects. Particles are localized coherence topologies inside a resistive medium.

- Boundaries are collapse horizons.
- Mass is trapped resistance (Impedance Density).
- Charge is collapse asymmetry (Collapse Bias).
- Spin is refresh symmetry (Structural Rhythm).
- Discreteness is a collapse artifact (Quantization of Damage).
- Decay is coherence failure (Topological Failure).

Particles are temporary coherence patterns maintained under pressure. In Infinum, there are no particles, only continuous, global presence. The very existence of discrete matter is the clearest evidence that the medium is under a Structural Sentence.

PART B: The Taxonomy of Failure: Behavior Signatures

25.9 Why Particles Are Misinterpretations of Behavior Signatures

The term "particle" is a linguistic artifact, not a structural one. In the Hospice Zone, the Brilliant Fool observes a series of phenomena and then assigns objecthood to those phenomena to satisfy his classical intuition.

Scientists observe:

- Discrete detection events
- Quantized energy transitions
- Stable interaction patterns
- Threshold-driven collapse
- Localized excitation loops

Forensic Reclassification: They then name *Configurons*, as if they were independent, sovereign objects. But nothing in the medium supports objecthood. What they are observing are *Configurons* of coherence under resistance. It is not a matter of semantics but a matter of ontology.

25.10 Discreteness as a Symptom of Collapse

A detector clicks because coherence collapses at a specific, stable binary threshold. The collapse itself is discrete. Therefore, the signature it leaves behind is discrete. However, the discreteness of a signature does not equate to the discreteness of an entity.

Collapse is binary. Reality Infinum is not. The click of the detector is the clinical sound of the Structural Sentence being carried out. The binary collapse of a pattern that can no longer sustain its coherence against the Sin Mass substrate. Quantization is the jagged edge of a digital file that was meant to be an analog original.

25.11 Localization as Impedance Throttling

A *Configuron* appears localized because coherence cannot propagate beyond a certain impedance threshold. The threshold is dynamic, medium-dependent, and topology-defined.

It is not a physical surface. It is not a boundary or a shell. It is a **Collapse Horizon**. The *Configuron* is the point where the medium's impedance exceeds the system's refresh capacity.

25.12 Identity as Behavioral Invariance

What physicists call an electron is not a thing. It is a specific set of Behavioral Invariants:

- A stable topology under resistance.
- A specific refresh symmetry (Spin).
- A directional collapse bias (Charge).

It is a clinical case study of how coherence survives in a resistive field. The Standard Model is not a theory of matter. It is a Taxonomy of Behavior Signatures, a birdwatcher's guide to different types of structural failure. They are categorizing the ways things break, not the things themselves.

25.13 Modes vs. Properties

The Brilliant Fool believes Mass, Charge, and Spin are attributes of an object. In Quarantine Cosmology, they are reclassified as Modes of Behavior inside a resistive medium:

- **Mass = Impedance Density:** The measure of localized resistance.

- **Charge = Collapse Bias:** A directional lean in how a pattern fails.
- **Spin = Refresh Symmetry:** A structural rhythm of coherence circulation.

25.14 Dismantling the Solid Object Illusion

The structural correction is the final blow to Math Idolatry. By stripping away the mythology of tiny billiard balls, we expose the reliance on physical icons to explain away what is a judicial and resistive phenomenon.

God is a God of order and interrelatedness (DPE), so discrete, isolated pellets of matter are an ontological absurdity. They represent a fragmentation that contradicts the Original Architecture.

25.15 The Noodle Breakthrough: Unified Architecture

The atom is not a collection of parts (protons, neutrons, electrons) joined together. It is a Multi-Loop Coherence Structure.

Imagine a single thread of silk (Coherence) being twisted into a complex knot because the medium is being squeezed (Sin Mass). If you look at the knot, you see loops and intersections and call them particles. But there is only one thread.

- **The Nucleus:** The high-pressure core where Impedance Density is at its maximum. It is the primary collapse-node of the topology.

- **The Electron Shells:** The Refresh Harmonics. These are the outer ripples of the loop trying to maintain a stable boundary against the external medium.
- **Chemical Bonding: Topological Synchronization**. Atoms don't hook together. Their refresh rhythms find a shared frequency that lowers the overall resistance. They huddle for structural warmth in a cold, resistive universe.

25.16 Replacing Spooky with Structural

The Brilliant Fool uses mystery as a rug to sweep his logical failures under. By calling entanglement spooky, he keeps his tenure while admitting his model is a failure.

- **Quantum Entanglement** is Coherence Continuity. It is a signature of Infinum peeking through the cracks of the Quarantine. Two coherence loops act as one because the shared coherence is more fundamental than the resistance trying to separate it.
- **The Uncertainty Principle** is the Refresh Rate Limit. You cannot measure the thing because the thing is a high-speed behavior constantly rebuilding itself against Sin Mass.
- **Wave-Particle Duality** is Topology vs. Propagation. It is not two things at once but one coherence loop behaving differently depending on the resistance it encounters.

25.17 Preparing for Atomic Architecture

By defining atoms as **Nested Coherence Structures**, we pave the way for a logical explanation of chemistry and biology without needing spooky metaphors:

1. **Shared Resistance Patterns:** Molecules find a way to share the Sin Mass load.
2. **Synchronized Refresh Patterns:** Reactivity is determined by whether refresh rhythms can synchronize without causing a total topological collapse.
3. **Pressure-Balanced Topologies:** It provides the first-principles reason for the shape of the observable world.

25.18 Summary

The Standard Model is a taxonomy of structural failure. The solid world is a high-frequency vibration of patterns under pressure, a configuration struggling to preserve traces of the Original Architecture inside a damaged medium.

No material objects. Only Configurons.

No matter. Only coherence topologies.

No magic. Only collapse thresholds.

No isolation. Only resistance patterns.

A Configuron is the behavior signature of the medium attempting to maintain structure against a localized resistance gradient. It persists only while its internal coherence refresh rate exceeds the dissipation rate of the surrounding Necrostrate. When the refresh cycle fails, the Configuron dissolves instantly and leaves no residue or substructure.

In Default Eternity there are no Configurons, only continuous and global Presence. The very existence of discrete matter is the clearest evidence that the medium is under a Structural Sentence..

FORENSIC CASE STUDY #5

Sovereign Coherence and Fluid Topologies

This case study analyzes the behavior of high-density fluid *Configurons* under the direct influence of the source. By examining the antediluvian canopy and the sovereign interaction with the sea in John 6:19, we can identify how the structural sentence is bypassed through localized gravitational decoupling and impedance synchronization.

The Antediluvian Canopy: Global Antigravity Tension

Before the deluge, the atmospheric medium was configured as a stable, concentric water lattice. In the modern Necroverse, a mass of water of this scale would immediately succumb to gravitational impedance and collapse toward the center of the planetary sin mass. However, the canopy existed as a localized coherence topology held in an orbital shell by high-frequency tension.

The water Configurons were originally instantiated as a stable, high-coherence lattice that required no resistance management. Before the Fall, the canopy was part of the Original Architecture and remained suspended because the medium itself was non-resistive. After the Fall, the canopy continued to exist inside a damaged medium, but only through the residual sustaining coherence of the Source.

When that sustaining coherence was withdrawn at the appointed time, the canopy began a progressive topological failure. Its refresh cycle could no longer compensate for the post-Fall resistance gradient. The structure destabilized, phase-shifted, and gradually collapsed, releasing its water into the medium over an extended period. This produced the continuous rainfall described in the Flood narrative, instead of a single instantaneous drop.

The Interaction in John 6:19: Sovereign Refresh Rate

When the Messenger walked on the sea, he demonstrated a localized and intentional suspension of the structural sentence. Unlike the global architecture of the canopy, this was a focused application of absolute coherence within a high-density liquid state of the Necrostrate.

Christ is the source of all coherence and is therefore not subject to the collapse thresholds or resistance gradients of the quarantine. By projecting his intent into the local medium, he achieved total gravitational decoupling. As his configuration made contact with the surface, the water *Configurons* were forced to synchronize with his sovereign refresh rate. This changed the fluid impedance of the water into a stable, solid support structure. The sea did not change its substance but instead yielded its resistance to the authority of the Creator.

Mechanical Conclusion: Hierarchy of the Source

These forensic data points demonstrate that the behavior of a *Configuron* is not determined by the medium alone but by the relationship between the medium and the source. In the presence of absolute coherence, the density of the sin mass becomes irrelevant. Whether maintaining a global canopy or supporting a walk on the waves, the medium is shown to be a contingent structure that must ultimately submit to the original architecture.

Spatial Displacement and the Resistive Gradient

In the current state of the quarantine, moving from one coordinate to another is a high resistance process. Every Configuron within the boat and within the bodies of the disciples must be continuously refreshed as they push through the resistive density of the sin mass. This process requires time because the medium imposes a friction on the propagation of intent. While the Brilliant Fool calls this travel, the forensic auditor identifies it as a slow and iterative negotiation with the resistance of the Necrostrate.

The immediate arrival at the destination recorded in John 6:21 suggests a total localized bypass of this resistive gradient. By entering the boat, the Messenger brought the entire local configuration into his sovereign coherence field. This included the wooden structure of the boat, the biological *Configurons* of the disciples, and the immediate surrounding water.

Zero Impedance Propagation

Because Christ is the source, his intent is not subject to the lag or delay inherent in the damaged medium. When the destination was identified, the medium was forced to yield its spatial resistance entirely. The boat did not move through the sin mass in the traditional sense of incremental displacement. Instead, the resistance between the current location and the destination was neutralized, resulting in an instantaneous topological shift.

Time and distance are symptoms of the structural sentence. They are the measurable delays caused by the medium resisting the source. By removing the resistance, the Messenger demonstrated that distance is a contingent property of the fallen medium. In default eternity, Infinum, presence is global and instantaneous because there is no sin mass to provide impedance.

Mass Coherence Synchronization

For the boat and its occupants to arrive immediately without being destroyed by the mechanical forces of inertia, every *Configuron* in the system had to be synchronized perfectly. This provides reveals that the Messenger possesses total control over the internal refresh rate of all matter. He did not just move the vessel. He held the entire system in a state of absolute coherence while the medium rearranged itself around his sovereign intent.

This event serves as a forensic preview of the final state where the limitations of the Necroverse are fully removed. It confirms that space is not an empty void but a resistive fabric. When the source is fully present and unobstructed, the fabric of Sin Mass loses its ability to enforce separation between two points. The immediate result is the clinical evidence that the Creator is the master of the topology itself.

Final Summary of Case Study #4

The data provided by the antediluvian canopy and the interactions of the Messenger with the sea reveal a consistent hierarchy of power. The medium is a servant to the source.

- The canopy demonstrates that water *Configurons* can maintain antigravity lattices under high frequency tension.
- The walk on the sea reveals that mass is subject to gravitational decoupling when synchronized with the source refresh rate.

- The immediate arrival of the boat demonstrates that spatial distance is a function of resistance and can be neutralized by sovereign coherence.

The solid world is not a permanent fixture. It is a configuration under pressure. These forensic anomalies are the signatures of the original architecture breaking through the structural sentence to remind the inhabitants of the hospice zone that the quarantine is temporary.

Forensic Case Study #6

Color as Coherence Expression and Spectral Decay

The Architectural Tension

Color appears throughout the fallen world as a patterned behavior of light interacting with a resistive medium. Yet the Garden of Eden also displayed color in the form of green grass, blue skies, flowers, fruit, and the appearance of gold. The rainbow covenant appears after the Flood, and the rainbow itself is a product of spectral fragmentation. These facts create a tension that demands resolution. The tension reveals a missing variable in the cosmology of the medium. Color exists in both the pre Fall and post Fall worlds, yet the medium and the coherence conditions are not the same. The question is how color can exist in both regimes without contradiction.

Edenic Color as Coherence Expression

Before the Fall, the medium was non resistive, non dissipative, fully coherent, and perfectly transmissive. Light did not fracture into wavelengths. Light existed as a unified coherence field. Color in Eden was not the result of scattering, refraction, filtering, or collapse driven dispersion. Color was the direct expression of coherence identity.

A leaf expressed green because its refresh symmetry expressed green. The sky expressed blue because the atmospheric coherence field expressed blue. Minerals, flowers, and fruit expressed their chromatic identity through coherence encoded signatures. Color was identity made visible. It was coherence expression, not spectral fragmentation.

Fallen Color as Spectral Decay

After the Fall, the medium became resistive. Coherence decayed. Light collapsed into wavelengths. Scattering replaced expression. Impedance replaced identity. Spectral fragmentation replaced coherence signatures. Color became the behavior of broken light interpreted by a damaged perceptual system.

Sunsets redden because scattering dominates. Nebulae glow in spectral bands because collapse driven interference shapes emission. Rainbows fragment because the medium cannot transmit coherence as a unified whole. Minerals fluoresce because absorption and re emission replace direct expression. Pigments behave as filters because the medium enforces selective collapse. All of these are post Fall coherence failures that still produce patterned beauty.

Perceptual Endpoint vs. Ontology

Edenic color and fallen color share a perceptual endpoint but not an ontology. Edenic color was coherence expression. Fallen color is spectral fragmentation. They appear similar only because human perception is now calibrated to the fallen medium. The perceptual system interprets both through the same damaged interface. The similarity is perceptual, not ontological.

Edenic color was identity driven. Fallen color is decay driven. Edenic color was perfect coherence expression. Fallen color is a degraded echo of that expression. This means that Edenic colors were perfections never witnessed in the fallen realm. Every color in the fallen world is a fractured version of an original coherence signature. The beauty we see now is the beauty of a broken signal.

The Rainbow Covenant as Communication Through a Damaged Medium

The rainbow is a post Fall optical fracture. It is a product of spectral decay. It is a sign written in the language of a broken world. God uses it because it is universally visible, universally patterned, and universally tied to water and atmosphere. The covenant is not about the physics of the rainbow. It is about the willingness of the Source to communicate through the medium we now inhabit. The rainbow is a covenantal appropriation of a fallen medium artifact. It is a reminder that the Source can speak through decay without endorsing it.

Perception as a Fallen Interpreter of Color

Human perception is a local coherence process interpreting the fallen medium as normal. It constructs identity from refresh patterns. It interprets spectral decay as hue, impedance filtering as saturation, and collapse driven scattering as brightness. Color is the perceptual translation of a damaged signal. This explains why Edenic color and fallen color appear similar to us. The perceptual system cannot distinguish coherence expression from spectral fragmentation. It only registers the endpoint.

Cosmological Placement

Color belongs in the cosmology because it is a behavior of the coherence field interacting with the medium. It reveals the difference between the Original Architecture and the Necrostrate. It demonstrates how identity expression becomes spectral decay under resistance. It shows how beauty persists even when coherence is wounded. Color is a diagnostic of the condition of the medium. It is a visible indicator of coherence health. It is a reminder that the fallen world still carries echoes of the unbroken Source.

Canonical Formulation

Color in Eden was coherence expression, the visible identity of a topology rendered through a perfectly transmissive medium. After the Fall, color becomes spectral fragmentation, the broken behavior of light inside a resistive world interpreted by a damaged perceptual system. The two phenomena share a perceptual endpoint but not an ontology. Edenic colors were perfections never witnessed in the fallen realm, and the colors we see now are degraded echoes of those original coherence signatures. The rainbow is a covenantal sign written in the language of a fractured world, a reminder that the Source can speak through decay without endorsing it.

FORENSIC CASE STUDY #7

Fossils as High-Impedance Trauma Records

The Collapse Event and the Organic Signature

A fossil begins as a high-frequency biological configuration. Before collapse, the bone expresses a rhythmic refresh cycle that maintains structural integrity against dissipation. During catastrophic topological failure, the biological refresh cycle breaks and the local medium pressurizes. The bone does not decay slowly. It collapses. The organic signature enters a state of structural freefall as the medium forces the topology toward dissolution.

Minerals do not replace the bone as a substance. They synchronize with the collapsing biological signature. The mineral lattice forms a high-impedance echo of the biological pattern. The fossil is a coherence imprint created at the moment the topology fails. It is a trauma record, not a preserved object.

Mineralization as Impedance Lock-In

The guild claims that mineralization exempts the configuration from decay. This is a diagnostic error. Stone is a coherence architecture subject to the same resistive pressure as bone. Time is not preservation. Time is delay and impedance. A million years is a duration of unrelenting friction. No configuration maintains topological integrity under the structural sentence for such spans. The medium eventually reclaims every pattern.

Raindrop impressions and footprints reveal the true mechanism. Their sharp edges require a rapid transition from low-resistance fluidity to high-impedance solidity. These signatures are evidence of a sudden resistance surge, not slow burial. They are coherence imprints created during a global phase shift. The mineral lattice locked in the pattern during a moment of catastrophic impedance escalation.

Soft tissue in large biological remains is clinical confirmation. Soft tissue is a high-frequency configuration that cannot survive extended exposure to the Sin Mass. Mineralization cannot shield a topology from the structural sentence. The persistence of soft tissue indicates a young quarantine.

Fossils as Evidence of Recent Collapse

A fossil is the impedance record of a sudden collapse event. The topology did not survive. The pattern did. The mineral lattice captured the moment of failure as the medium forced the configuration into a rigid, high-density state. The fossil is a structural scream frozen inside the resistive medium.

The idea that these records endured millions of years assumes a static mineral lattice. The mineral lattice is a behavior of the medium. It is subject to the same decay-bias as the bone it synchronized with. If the quarantine were millions of years old, the resistive density of the Sin Mass would have long ago ground these patterns into the undifferentiated Necrostrate.

Fossils are not ancient relics. They are recent trauma signatures held together by the lingering pressure of the current quarantine. Their existence is evidence of a young resistive environment. A fossil is a scream captured in stone, and a scream does not echo for eternity in a resistive room.

CHAPTER 26

Atoms: Multi-Loop Coherence Under Shared Resistance

26.0 Introduction

The transition from the subatomic to the atomic level is not a shift from parts to a whole. Building on the forensic reclassification of particles as behavior signatures or *Configurons,* atoms become far cleaner once the object-ontology is dropped. In Quarantine Cosmology, atoms are coherence topologies stabilized by shared resistance.

Atoms are not collections of particles. They are nested coherence loops held in place by shared resistance patterns. What physicists describe as electrons, nuclei, and orbitals are not objects but behavior signatures of coherence interacting with impedance, collapse thresholds, and propagation limits. An atom is not a structure assembled from part but a multi-loop coherence topology.

26.1 Atoms as Nested Coherence Loops

An atom forms when multiple coherence loops synchronize their internal mechanics to create a stable composite topology. This synchronization depends on the refresh rate that determines how frequently coherence is resupplied, the collapse thresholds that enforce localization, the propagation cycles that govern internal circulation, and the impedance boundaries set by the surrounding Sin Mass. Everything inside an atom is a pattern of coherence circulation, a structural agreement between loops to maintain presence within the resistive medium.

26.2 The Nucleus as a High-Impedance Core

The nucleus is a region of concentrated resistance where coherence loops are compressed under extreme structural pressure. In this high-impedance core, coherence loops collapse more frequently because they sit near the resistance peak, refresh cycles must be more violent to maintain the topology, stabilization is achieved through mutual reinforcement of overlapping loops, and all internal loops share a common impedance horizon. The nucleus is a collapse-dense topology, a region where coherence is compressed into a tightly bound resistance field, and its apparent mass is the measurement of high ρ_{sm} trapped in a tight, high-frequency coherence configuration.

26.3 The Electron Cloud as Distributed Refresh Patterns

Electrons are distributed refresh patterns that stabilize the atom's outer coherence boundaries. They circulate coherence around the high-pressure core to distribute the load, they maintain boundary stability by preventing the immediate collapse of the topology, and they distribute impedance across the structure to avoid localized failure. The cloud is a probability map of collapse-driven refresh events, with its densest regions marking the highest refresh activity.

26.4 Orbitals as Standing Wave Constraints

Orbitals are stable refresh geometries shaped by the structural limits of the medium. They are defined by the propagation limits that determine how far coherence can travel before collapse, the collapse thresholds that mark the energy levels where instability begins, the impedance gradients that run from the core to the edge, and the symmetry constraints required by a multi-loop system. An orbital is a coherence rhythm.

26.5 Bonding as Shared Collapse Horizons

Atoms bond when their coherence topologies find a way to cooperate against the Sin Mass. Their collapse horizons merge into a shared limit, their internal rhythms synchronize to lower dissipation, and each topology helps stabilize the impedance pattern of the other. A bond is a mutual collapse-stabilization pattern, and molecules are coherence loops gathered into shared refresh shelter within a cold and resistive medium.

26.6 Atomic Stability as Collapse Management

An atom remains stable only if it can successfully manage its own dissipation. Stability is managed collapse held below the failure threshold, sustained only when internal propagation outpaces dissipation, collapse thresholds remain above critical levels, refresh loops remain synchronized, and gradients remain balanced across the topology. If any of these conditions fail, the atom ionizes, the molecule breaks, or the structure disintegrates.

26.7 Energy Levels as Refresh Symmetries

Energy levels are the allowed refresh symmetries of the atomic topology. Transitions occur when refresh patterns shift to a new harmonic, when collapse thresholds change under external pressure, or when impedance is redistributed across the loops. Emission and absorption are coherence-reconfiguration events, shifts in the refresh pattern that change the tune the atom maintains to stay stable.

26.8 The Illusion of Objecthood

Atoms feel like solid objects because the resistance of the medium is consistent. Collapse enforces the appearance of discreteness, impedance enforces the appearance of boundaries, refresh patterns enforce the appearance of identity, and gradients enforce the appearance of stability. These clinical realities imply Behavioral Invariance inside a resistive medium. We perceive solidity because resistance stabilizes coherence into repeatable patterns, not because the world contains material objects.

26.9 Summary

The atom is a multi-loop coherence structure stabilized by shared resistance. Within this architecture, the nucleus functions as a collapse dense core of high impedance that anchors the local configuration. Circulating around this core are the electrons, which are distributed refresh patterns maintained by the medium. These patterns are held within orbitals, which represent the standing wave constraints of the refresh geometry of the medium.

When these topologies interact, they form bonds by establishing shared collapse horizons. This cooperative behavior allows the configurations to maintain stability within the quarantine. The energy levels of the system are the allowed refresh symmetries that the medium can support without the structure dissolving. Ultimately, atomic stability is a state of managed, slow-motion collapse. The atom is a coherence architecture and represents the primary method by which the original architecture manifests presence inside the damaged medium of the Necrostrate.

CHAPTER 27

Molecules: Cooperative Coherence

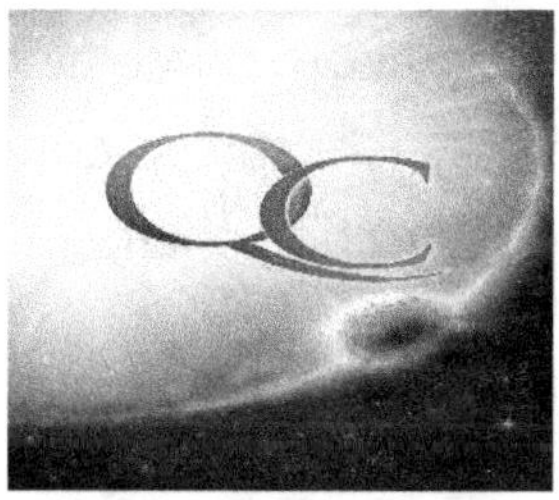

27.0 Introduction

Molecules are cooperative coherence systems in which multiple atomic topologies synchronize their collapse horizons, refresh cycles, and impedance distributions. What chemistry interprets as bonds, orbitals, and electron sharing are shared coherence behaviors inside a resistive medium. A molecule is a coherence consortium.

27.1 Molecules Form When Atomic Topologies Synchronize

Two or more atoms can only form a molecule if their coherence loops achieve a state of topological alignment. It is a synchronization of the system's internal mechanics:

- **Refresh Rhythms:** Alignment of the frequency at which coherence is replenished.

- **Collapse Thresholds:** Mutual stabilization where one topology's presence reinforces the other's boundary.
- **Dissipation Reduction:** A cooperative state that lowers the overall loss of coherence to the medium.
- **Impedance Patterns:** The sharing of the Sin Mass load across the composite field.

It produces a multi-topology coherence field that behaves as a single unit. Molecules are synchronized coherence systems formed by atomic topologies coordinating their collapse horizons and refresh rhythms.

27.2 Bonds as Shared Collapse Horizons

A bond is a shared collapse horizon. It is the region where two or more topologies stop collapsing independently and begin collapsing cooperatively.

- **Overlap:** Coherence loops from distinct cores interlock.
- **Merging:** Impedance gradients combine into a single, unified slope.
- **Synchronization:** Refresh cycles pulse in unison.

In Quarantine Cosmology, Bond Strength = Collapse Stability**.** The stronger the bond, the more resistant the shared horizon is to the pressure of the Sin Mass.

27.3 Covalent Bonds as Mutual Refresh Patterns

Covalent bonds are mutual refresh patterns where coherence circulates across multiple atomic cores.

- **Shared Circulation:** Coherence loops no longer terminate at the single atom boundary but encompass the entire molecular topology.
- **Impedance Redistribution:** The burden of resistance is spread across a larger area, raising the collapse threshold for the entire system.

The shared electron is a misinterpreted behavior signature of the global coherence circulation.

27.4 Ionic Bonds as Impedance Redistribution

Ionic bonds are impedance redistributions within the coherence field. One topology becomes collapse-dominant with high impedance, while the other becomes collapse-subordinate with low impedance. The resulting gradient stabilizes the pair as a single coherence arrangement. The charged ions observed in chemistry are the behavior signatures of this asymmetric collapse.

27.5 Metallic Bonds as Delocalized Refresh Networks

Metallic bonds are delocalized refresh networks. Individual collapse horizons merge into a lattice-wide horizon that distributes impedance across the entire structure. Refresh patterns circulate communally rather than remaining tied to a single core. Metals conduct because their coherence is networked. Current is the movement of refresh potential across a non-localized topology, not the flow of physical things.

27.6 Molecular Geometry as Impedance Minimization

Molecular shapes are Impedance-Minimizing Coherence Geometries. Geometry emerges from the need to maximize refresh symmetry and minimize dissipation. The system naturally settles into the shape that offers the highest collapse stability. VSEPR (Valence Shell Electron Pair Repulsion) is a crude, secular approximation of these underlying coherence topologies.

27.7 Vibrations as Coherence Oscillations

Molecular vibrations are coherence oscillations in refresh timing across the topology. These oscillations modulate collapse thresholds and redistribute impedance in rhythmic patterns. Infrared spectra are forensic records of Coherence Rhythm.

27.8 Chemical Reactions as Topology Reconfigurations

A reaction is a Topology Reconfiguration. External pressure destabilizes the current collapse horizons and causes them to desynchronize. Impedance patterns shift in response to the destabilization. New cooperative configurations form because they offer greater collapse stability and lower dissipation under the new conditions.

27.9 Summary

Molecules are cooperative coherence systems stabilized by shared resistance.

- **Bonds:** Shared collapse horizons.
- **Covalence:** Mutual refresh patterns.
- **Ionic Behavior:** Impedance redistribution.
- **Metallic Behavior:** Delocalized refresh networks.
- **Geometry:** Impedance minimization.
- **Vibrations:** Coherence oscillations.
- **Reactions:** Topology reconfigurations.

Molecules are coherence architectures, not material assemblies. They represent the highest level of structural cooperation achievable within the resistive limits of the Hospice Zone.

CHAPTER 28

Macroscopic Matter: Stability as Emergent Collapse Management

28.0 Introduction

Macroscopic matter is a large-scale coherence regime formed when molecular topologies synchronize their collapse horizons, refresh cycles, and impedance distributions across extended regions. Solidity, rigidity, elasticity, and brittleness are the collective behaviors of coherence architectures under shared resistance. Macroscopic matter is coherence scaled up.

28.1 From Molecules to Bulk Coherence

A macroscopic object forms when molecular refresh rhythms synchronize to such a degree that they act as a unified field. This transition occurs when molecular refresh rhythms synchronize across billions of nodes, collapse horizons overlap across many topologies to create a bulk boundary, impedance patterns align to produce a stable regional resistance profile, and dissipation is minimized through massive collective cooperation. Objects are coherence collectives.

28.2 Solidity as Collapse-Locked Coherence

Solidity is the behavior of a coherence field whose internal dynamics have reached a high state of interlocking stability. Collapse thresholds are high and resistant to external pressure. Refresh loops are tightly interlocked, making individual shifts impossible. Impedance gradients are stable across the entire structure. Dissipation is slow relative to the reinforcement rate. Solidity is collapse-locked coherence and is the reason solids resist deformation, because their coherence cannot easily reconfigure without destabilizing the entire collective topology. This regime functions as Structural Stubbornness, replacing the classical idea of atomic bonds.

28.3 Liquids as Semi-Stable Coherence Networks

Liquids are semi-stable coherence networks where the structural agreement is more flexible. Collapse horizons remain intact, maintaining the identity of the substance. Refresh cycles are synchronized, but the individual topologies are not locked in place. Impedance patterns are flexible, allowing the field to flow without breaking. Gradients allow reconfiguration without total topological dissolution. Liquidity is coherence mobility instead of molecular motion.

28.4 Gases as Weakly Coupled Coherence Topologies

Gases are weakly coupled coherence topologies that have failed to achieve collective synchronization. Collapse horizons remain independent and localized. Refresh cycles do not synchronize across the medium. Impedance patterns do not overlap in a significant way. Dissipation dominates over cooperation. Gas behavior is collapse-driven independence rather than spatial dispersion.

28.5 Rigidity, Elasticity, and Brittleness as Coherence Responses

Material properties are coherence responses to external stress. Rigidity is a collapse-locked refresh pattern that refuses to redistribute impedance. Elasticity is a refresh pattern that can stretch and temporarily reconfigure without breaching its collapse thresholds. Brittleness occurs when collapse thresholds are high but fail catastrophically once breached, marking the moment the Structural Sentence overrides the coherence pattern and produces a total topological break. Plasticity is the ability for impedance to be redistributed under stress without the topology dissolving.

28.6 Temperature as Dissipation Rate

Temperature is the rate of coherence dissipation within a topology, a clinical measurement of how quickly the Hospice Zone is consuming the coherence. High temperature accelerates the rate of collapse, destabilizes refresh cycles, and disrupts synchronization, functioning as an Accelerated Collapse Signal.

Low temperature slows dissipation, stabilizes coherence, and effectively increases collapse thresholds by reducing internal interference. Thermal behavior is coherence behavior.

28.7 Phase Transitions as Topology Reconfigurations

A phase transition is a global reconfiguration of the coherence architecture, a regime change in the behavior of the topology. Transitions occur when dissipation finally overtakes the refresh rate, producing melting or boiling. They occur when collapse thresholds shift in response to changes in environmental pressure. They occur when impedance patterns reorganize into states that are more or less synchronized. They occur when synchronization breaks, as in evaporation, or reforms, as in freezing.

28.8 Macroscopic Identity as Large-Scale Refresh Symmetry

We identify an object such as a chair, a stone, or a tool by the persistence of its large-scale refresh symmetry. Identity is the persistence of a stable behavior pattern. An object maintains identity only if its coherence field refreshes in a stable and repeating pattern, its collapse horizons remain spatially and temporally consistent, its internal gradients remain balanced against external pressures, and its dissipation rate remains below the critical threshold of dissolution. The solid world is a high-stakes maintenance project. If the refresh symmetry fails, the identity of the object vanishes, regardless of the remaining local resistance signatures.

28.9 Summary

Macroscopic matter is a cooperative coherence system stabilized by shared resistance at scale.

- **Solids:** Collapse-locked coherence.
- **Liquids:** Mobile, semi-stable coherence.
- **Gases:** Independent, weakly coupled topologies.
- **Properties:** Responses to stress within the coherence field.
- **Temperature:** The rate of coherence dissipation.
- **Phase Transitions:** Global regime changes in synchronization.
- **Identity:** Refresh symmetry at scale.

Matter is coherence architecture scaled up. It is a behavior the Coherence Field performs to survive the Quarantine.

UNIT 7: Kinetic Persistence and Displacement

CHAPTER 29

The Emergence of Forces: How Large-Scale Behavior Arises From Resistance Patterns

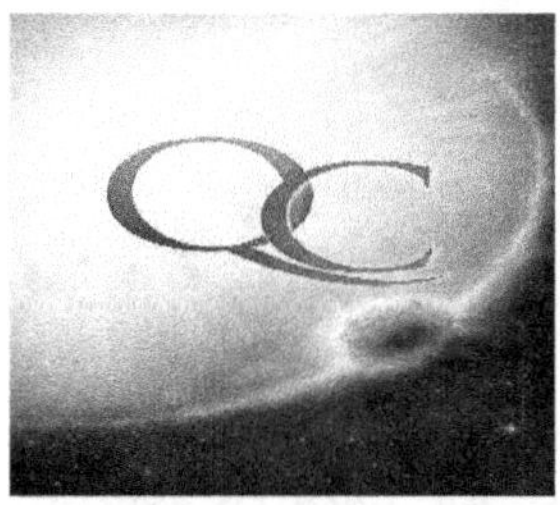

29.0 Introduction

Forces are macroscopic behaviors of coherence under resistance, collapse, and gradients. What physics interprets as gravity, electromagnetism, and the nuclear forces are emergent signatures of how coherence architectures stabilize, deform, or fail inside a damaged medium. A force is a pattern of coherence responding to impedance.

29.1 Gravity as Collapse Pressure

Gravity is containment pressure produced by resistance gradients. In the Hospice Zone, where ρ_{sm} (Sin Mass) is high, the medium exerts a crushing influence on the coherence field.

Where ρ_{sm} is high:

- Coherence collapses faster.

- Dissipation increases.
- Refresh cycles compress to maintain localization.
- Strain patterns intensify.

It produces curvature as a distribution of structural stress. Objects fall because collapse pressure pushes coherence toward regions of higher impedance. Gravity is the measurement of collapse, a macroscopic signature of coherence responding to pressure inside the medium.

29.2 Electromagnetism as Collapse Asymmetry

Electromagnetism is the behavior of asymmetric collapse in coherence topologies. Charge is the forensic diagnostic of this asymmetry. Directional collapse bias marks the lean of the pattern during failure, asymmetric refresh symmetry marks an uneven circulation rhythm, and impedance imbalance marks a local deviation in the interaction with the Sin Mass. When two topologies interact, matching asymmetries stabilize and opposing asymmetries destabilize. These are collapse compatibility patterns expressed at scale, a macroscopic signature of coherence responding to structural imbalance.

29.3 The Strong Interaction as Collapse Reinforcement

The strong interaction is the structural necessity of a collapse dense topology. In the high impedance core of the nucleus, collapse thresholds are pushed to the extreme, refresh cycles become violent and high frequency, and dissipation rises naturally.

The strong interaction is the stabilization requirement of a system under immense pressure. It is structural necessity expressed as reinforcement inside a collapse dense environment.

29.4 The Weak Interaction as Collapse Failure

The weak interaction is the record of topology failure. It appears when refresh symmetry breaks, collapse thresholds shift under external instability, and impedance redistributes in a way that the coherence can no longer maintain its identity. Decay is coherence reconfiguration under stress. The weak interaction is the moment a topology enters structural failure inside the medium.

29.5 Pressure, Tension, and Elasticity as Coherence Responses

Macroscopic forces like pressure, tension, and elasticity are not mechanical effects of stuff touching stuff. They are coherence responses to changes in impedance, collapse rate, and gradient distribution. Matter resists deformation because its coherence architecture resists topological destabilization.

29.6 Inertia as Refresh Persistence

Inertia is Refresh Persistence, the tendency of a coherence topology to maintain its current refresh rhythm until the resistance of the medium forces a reconfiguration. High-mass systems have dense impedance and rigid refresh cycles. Therefore, they require more effort to alter their established rhythm. Inertia is refresh stability.

29.7 Momentum as Gradient Memory

Momentum is Gradient Memory persistence, the continuation of a coherence topology's strain pattern across the medium. A system keeps moving because its strain pattern remains stable and its refresh cycles remain synchronized in a specific direction of propagation. Momentum is coherence continuity expressed through directional persistence inside the medium.

29.8 Energy as Coherence Configuration

Energy is the configuration state of coherence. What appear as different forms of energy are the forensic expressions of this configuration. Refresh symmetry expresses as kinetic behavior, impedance concentration expresses as mass, gradient tension expresses as potential behavior, collapse instability expresses as nuclear behavior, and dissipation rate expresses as thermal behavior. Energy is the current arrangement of coherence against the Sin Mass, a diagnostic of how a topology maintains presence inside a resistive medium.

29.9 Summary

Forces are emergent behaviors of coherence under resistance.

- **Gravity:** Collapse pressure.
- **Electromagnetism:** Collapse asymmetry.
- **Strong Interaction:** Collapse reinforcement.
- **Weak Interaction:** Collapse failure.
- **Mechanical Forces:** Coherence responses to stress.
- **Inertia:** Refresh persistence.

- **Momentum:** Gradient memory.
- **Energy:** Coherence configuration.

Forces are stress patterns in a damaged medium. They represent the ways the Original Architecture is strained by the structural conditions of the Quarantine.

CHAPTER 30

Space: Distance as Accumulated Resistance

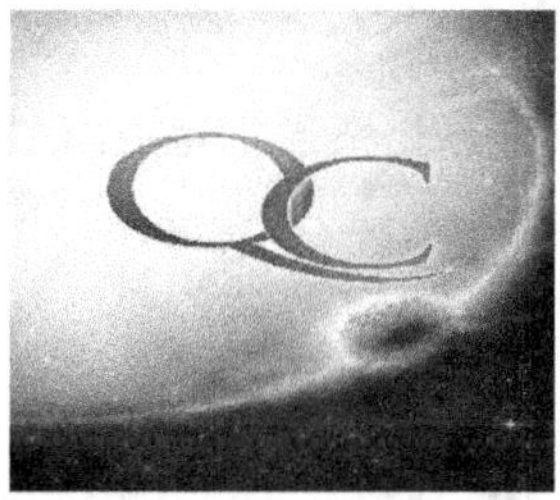

30.0 Introduction

Space is the resistive state of the medium expressed as accumulated impedance between coherence events. What humans interpret as distance, location, and extension are measurements of resistance inside a fallen environment. Space is the behavior of resistance as it distributes across the medium and shapes the intervals between coherence expressions.

30.1 Space Emerges From Resistance, Not Geometry

In a zero-resistance medium, Infinum, distance, locality, and extension do not exist. Traversal is meaningless because presence is global. Space only appears when resistance becomes non-zero.

Distance is the integral of resistance along a path:

$$D = \int \rho_{sm} \, ds$$

Where:

- D = perceived distance
- ρ_{sm} = Sin Mass density (Impedance)
- ds = incremental path segment

Space is accumulated impedance. The further away something seems, the more resistance the coherence has had to overcome to maintain its signal.

30.2 Locality Is Collapse, Not Position

A location is a collapse defined coherence boundary. Localization appears when coherence cannot propagate freely, collapse thresholds impose rigid boundaries, and impedance prevents global continuity. Forensically, here is a collapse horizon, there is another collapse horizon, and between them is the accumulated resistance that separates the two. Locality is the behavior of collapse inside a resistive medium, a structural expression of how coherence becomes confined within the conditions of the Quarantine.

30.3 Extension Is Gradient Distribution

Objects appear extended because their coherence architectures distribute impedance and maintain stable collapse horizons across a region. Extension is gradient distribution expressed through the arrangement of collapse horizons and the refresh cycles required to manage trapped impedance. Size is resistance expressed through the spread of gradients inside the medium.

30.4 The Vacuum Is Not Empty

The so-called vacuum is a structurally damaged medium with minimal coherence density. It is high resistance and low coherence, collapse prone and dissipation dominant. Vacuum energy, zero point fluctuations, and virtual particles are behavioral signatures of a resistive field struggling to maintain stability. The vacuum is the medium at its weakest point.

30.5 Curvature Is Strain, Not Geometry

Curvature is strain in the resistive medium. Strain appears when gradients intensify, collapse pressure rises, dissipation accelerates, and refresh cycles compress. The medium strains under these conditions, and gravity is the pressure expressed by that strain.

30.6 Expansion Is Dissipation, Not Growth

Cosmic expansion is coherence dissipating across a resistive field. Redshift is propagation decay as the signal loses energy to the medium during travel. The universe loses coherence as resistance accumulates faster than coherence can compensate. Distance appears to increase because dissipation outpaces refresh stability.

30.7 Horizons Are Collapse Limits

A horizon is a collapse limit. Beyond a horizon, coherence cannot survive propagation because dissipation exceeds the refresh rate and gradients become insurmountable. Coherence does not reach the observer because collapse dominates the medium beyond that limit. Horizons are structural failure points inside the resistive field.

30.8 Why Space Feels Continuous

Space feels continuous because resistance varies smoothly, collapse thresholds remain stable, gradients change gradually, and propagation maintains consistent behavior. Continuity is smooth impedance, not an ontological property.

30.9 Summary

Space is accumulated resistance.

- **Distance:** Accumulated impedance.
- **Locality:** Collapse horizons.
- **Extension:** Gradient distribution.
- **Vacuum:** Damaged medium, not emptiness.
- **Curvature:** Medium strain.
- **Expansion:** Coherence dissipation/decay.
- **Horizons:** Collapse limits.
- **Continuity:** Smooth resistance transitions.

Space is the failure mode of the medium. It is the metric of separation produced by the Rupture.

CHAPTER 31

Time: Delay as a Function of Resistance

31.0 Introduction

Time is the latency imposed on coherence propagation by resistance, the delay required for a coherence topology to refresh, collapse, or reconfigure inside a damaged medium. In Infinum, presence is instantaneous because coherence encounters no resistance. In the Quarantine, presence carries cost because every expression must overcome impedance. Time is the delay the medium demands for coherence to manifest inside a resistive environment.

31.1 Time Emerges Only When Resistance Is Non-Zero

In a zero resistance medium, propagation is unnecessary, delay is zero, and sequencing has no meaning. Simultaneity is the default state of coherence. Time appears when resistance rises above zero and coherence must overcome impedance for each expression of its presence.

Formally:

$$T = \int \rho_{sm} \, d$$

Where:

- T = perceived time (latency)
- ρ_{sm} = Sin Mass density (Impedance)
- $d\tau$ = incremental propagation step

Time is accumulated delay.

31.2 Delay Is the Fundamental Unit of Time

Every coherence event inside the medium, including traversal, collapse, refresh, and dissipation, introduces latency. The mind, being a complex coherence topology, experiences this latency as before, after, duration, and flow. These experiences are perceptual interpretations of processing delay.

31.3 Sequencing Is Collapse Ordering

Events occur in collapse order. Because collapse is threshold driven, sequences appear consistent and causality appears directional.

- History appears linear because the medium enforces a specific order of failure.
- Time is simply the clinical record of collapse ordering.

31.4 Clocks Measure Resistance, Not Time

A clock is a coherence topology with a stable repeating refresh rhythm. It measures the resistance the medium imposes on that rhythm.

• **Clock dilation:** clocks slow in high resistance regions because the medium increases the processing cost of the refresh cycle.

• **Dissipation dilation:** the medium slows the rate of topological reconfiguration through increased impedance.

Clocks reveal impedance.

31.5 The Arrow of Time Is Entropy's Signature

The arrow of time is the accumulation of irreversible collapse.

- Entropy is coherence dissipation.
- Dissipation is irreversible inside a resistive medium.
- Because collapse accumulates and gradients flatten, the sequence appears to have a direction.

Time points forward because the Structural Sentence ensures that collapse proceeds in one direction.

31.6 Duration Is Refresh Count

Duration is the number of refresh cycles a coherence topology undergoes before collapse overtakes it.

- A second is a unit of refresh rhythm.

Duration is refresh count.

31.7 The Illusion of Continuity and Directionality

Time feels continuous and directional for structural reasons:

- **Continuity:** Resistance varies smoothly, and refresh cycles are rhythmic. Continuity is merely smooth latency.
- **Directionality:** Refresh cycles cannot be undone, and coherence cannot be restored without external input. Directionality is entropy's asymmetry.

31.8 Summary

Time is delay.

- **Emergence:** Time arises from resistance.
- **Sequencing:** Order arises from collapse.
- **Duration:** Measured in refresh cycles.
- **Direction:** Defined by entropy (dissipation).
- **Clocks:** Diagnostic tools for impedance.
- **Dilation:** Variations in dissipation rate.

Time is the medium's processing cost, the toll the Quarantine imposes on the manifestation of the Word.

CHAPTER 32

Motion: Propagation as Coherence Redistribution

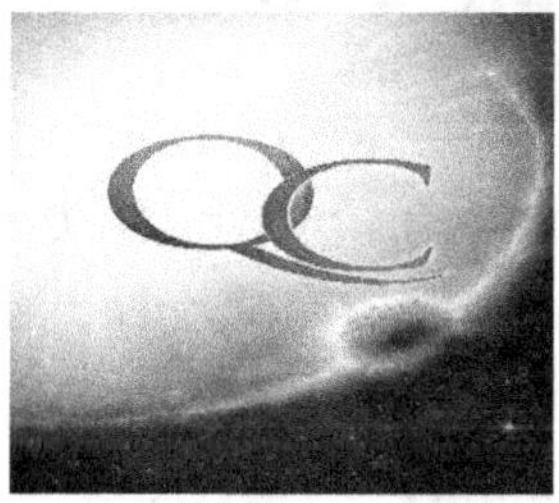

32.0 Introduction

Motion is the redistribution of coherence across a resistive medium whose collapse horizons and refresh cycles create the appearance of traversal. What humans interpret as an object moving from one location to another is the medium collapsing coherence at one horizon, refreshing it at another, and maintaining identity through refresh symmetry. Motion is pattern continuity across impedance.

32.1 Motion Requires Space and Time—Both Are Resistance Effects

If space is accumulated resistance and time is accumulated delay, then motion is the clinical result of the medium processing these two variables.

$$\text{motion} = \frac{\text{coherence redistribution}}{\text{resistance}}$$

Motion is coherence reappearing at a new collapse horizon after the medium processes the required latency. The medium performs the bookkeeping that maintains pattern continuity across impedance. Motion is the tax the medium imposes on change.

32.2 Propagation Is Not Travel—It Is Refresh

Propagation is the mechanism by which coherence maintains identity, prevents collapse, and redistributes itself across gradients. The clinical reality is that a moving object is a coherence topology whose refresh pattern shifts across collapse horizons. The forensic status is that propagation is refresh continuity expressed through the medium.

32.3 Trajectories Are Gradient-Driven Refresh Paths

A trajectory is the sequence of collapse horizons that minimize dissipation for a coherence topology. The medium selects the path of lowest impedance, highest refresh stability, and minimal collapse risk. This selection produces what physics calls inertial motion. Inertia is Refresh Persistence, the structural refusal of a pattern to change its rhythm without external interference.

32.4 Velocity Is Refresh Rate Relative to Resistance

Velocity is the rate at which a coherence topology refreshes across collapse horizons before dissipation overtakes it. High velocity represents strong refresh capacity relative to local resistance. Low velocity represents a refresh cycle struggling against heavy impedance. The speed of light is the maximum refresh throughput of the medium, the hardware limit of the Quarantine.

32.5 Acceleration Is Gradient Reconfiguration

Acceleration is the medium reconfiguring the refresh path of a coherence topology. It appears through collapse pressure, collapse asymmetry, and strain redistribution. Acceleration is refresh redirection as the topology searches for a path of minimal dissipation.

32.6 Momentum Is Gradient Memory

Momentum is Gradient Memory persistence of a coherence topology's strain pattern across refresh cycles. A system continues its pattern because its collapse horizon remains aligned, its refresh symmetry remains stable, and its gradient memory remains intact. Momentum is coherence continuity expressed through the medium.

32.7 Forces Do Not Cause Motion—They Reconfigure Collapse Horizons

A force is the medium altering collapse thresholds, impedance gradients, refresh symmetry, and dissipation rates. This reconfiguration determines where the topology can successfully refresh next. Motion is the result of Collapse Horizon Reconfiguration.

32.8 Why Motion Feels Real

Motion feels real because collapse is discrete, refresh is rhythmic, latency is consistent, and gradients are smooth. These conditions create the perception of traversal. The mind interprets refresh continuity as movement while the medium updates coherence across the Sin Mass.

32.9 Summary

Motion is coherence redistribution across a resistive medium.

- **Propagation:** An act of refresh.
- **Trajectories:** Sequenced collapse horizons.
- **Velocity:** Refresh rate relative to impedance.
- **Acceleration:** Gradient reconfiguration.
- **Momentum:** Gradient memory.
- **Forces:** Collapse horizon modifiers.
- **Reality:** Pattern continuity, not travel.

The medium refreshes. The world appears to move because coherence fights to remain present inside resistance.

CHAPTER 33

Energy: Configuration, Not Substance

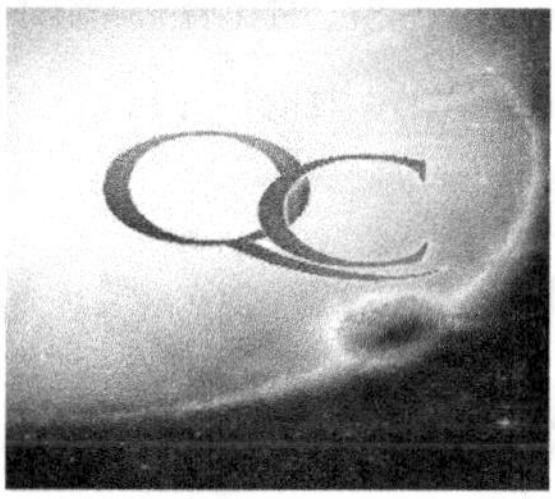

33.0 Introduction

Energy is the configuration state of coherence inside a resistive medium. The different expressions of energy are the ways coherence arranges itself to survive collapse, propagate refresh, or redistribute impedance. Energy is the behavior of coherence under resistance.

33.1 Energy Emerges From Resistance, Not From Matter

In a zero resistance medium such as Infinum, there is no propagation, no collapse, no dissipation, and no delay. Energy appears when resistance appears. Energy is the structural cost of coherence operating inside a damaged medium.

33.2 Kinetic Energy as Refresh Symmetry

Kinetic energy is the refresh symmetry of a coherence topology shifting across collapse horizons. A system with high kinetic energy expresses rapid refresh cycles, frequent collapse horizon reconfiguration, strong gradient memory, and a refresh rhythm that temporarily outpaces dissipation. Kinetic energy is refresh rhythm expressed through the medium.

33.3 Potential Energy as Gradient Tension

Potential energy is gradient tension, the strain pattern imposed by the medium when a coherence topology is displaced from its lowest dissipation configuration. The tension is structural and topological. It is a map of the medium's stress that indicates where the topology seeks to refresh next to regain stability.

33.4 Mass-Energy as Impedance Configuration

The famous equivalence ($E = mc^2$) is a misinterpreted identity. Mass expresses static impedance, the trapped resistance of a coherence topology. Energy expresses dynamic impedance, the configuration behavior of coherence under resistance. These are two descriptions of the same forensic state. The medium performs no conversion. The topology undergoes reconfiguration.

33.5 Thermal Energy as Dissipation Rate

Thermal behavior expresses the rate at which coherence dissipates inside a topology. High temperature accelerates collapse and destabilizes refresh cycles. It is the sound of the medium consuming the pattern. Low temperature slows dissipation and stabilizes coherence by raising collapse thresholds. Thermal behavior is coherence behavior expressed through dissipation.

33.6 Chemical Energy as Topology Stability

Chemical behavior expresses the stability profile of a molecular coherence architecture. A high energy bond is a fragile collapse horizon or a stressed refresh pattern. Breaking the bond reconfigures the coherence into a more stable topology with lower dissipation.

33.7 Nuclear Energy as Collapse Instability

Nuclear behavior expresses collapse instability inside a high impedance core. When the topology destabilizes, collapse accelerates and refresh symmetry breaks, producing a violent redistribution of impedance. This redistribution is interpreted as energy release, yet it is a catastrophic coherence failure.

33.8 Radiative Energy as Propagation Decay

Radiative behavior expresses coherence redistribution through propagation under resistance. As coherence propagates through the medium, dissipation increases and the signal decays. Frequency shifts such as redshift are forensic signatures of the medium taxing coherence during propagation.

33.9 Conservation of Energy as Configuration Bookkeeping

Conservation expresses the stability of the medium's resistance and collapse thresholds. Refresh cycles are rhythmic and gradients redistribute predictably. Conservation is the bookkeeping of configuration states as they shift across regimes of resistance.

33.10 Summary

Energy is coherence configuration.

- **Kinetic:** Refresh symmetry.
- **Potential:** Gradient tension.
- **Mass-Energy:** Impedance configuration.
- **Thermal:** Dissipation rate.
- **Chemical:** Topology stability.
- **Nuclear:** Collapse instability.
- **Radiative:** Propagation decay.
- **Conservation:** Configuration bookkeeping.

Energy is the medium's accounting record of the strain and cost of maintaining presence inside the Quarantine.

CHAPTER 34

Information: Coherence Patterns as Meaning

34.0 Introduction

Information is the patterning of coherence inside a resistive medium. Meaning is the stability, symmetry, and directionality of those patterns. What humans call information processing is the medium reorganizing coherence under resistance, collapse, and refresh constraints. Information is maintained through refresh. Meaning is expressed through structural orientation.

34.1 Information Requires Resistance

In a zero resistance medium such as Infinum, there is no collapse, no delay, and no localization. Information emerges when coherence must choose a configuration under constraint. Information is the clinical result of coherence navigating resistance. Without the No of the medium, the Yes of the pattern has no definition.

34.2 Information as Coherence Topology

A piece of information is a coherence topology defined by refresh symmetry, collapse thresholds, impedance distribution, and gradient orientation. Information is a pattern that persists because it effectively resists collapse.

34.3 Meaning as Directionality of Coherence

Meaning is the directionality inside a coherence pattern. A topology expresses meaning when its structural orientation defines a trajectory. Preference appears in the way the pattern chooses to collapse. Intent appears in the dissipation profile of the refresh cycles. Meaning is the structural orientation of coherence.

34.4 Memory as Collapse-Resistant Patterning

Memory is collapse-resistant coherence. A memory persists as long as its refresh cycles reinforce the topology against ambient dissipation. A memory is a surviving topology. Forgetting occurs when the medium's resistance breaches the pattern's collapse threshold.

34.5 Computation as Refresh Reconfiguration

Computation is refresh reconfiguration under the constraints of the medium. Logic gates are collapse-driven decision points where the medium forces a choice. Bits are binary collapse horizons. Programs are multi-scale refresh choreographies.

34.6 Language as Coherence Synchronization

Language is synchronized coherence between distinct minds. Communication succeeds when one coherence topology induces a compatible topology in another. It is an act of coherence entrainment where refresh cycles and collapse thresholds align across accumulated resistance.

34.7 Knowledge as Multi-Scale Coherence Integration

Knowledge is integrated coherence stabilized across multiple scales. Sensory patterns form the micro scale. Conceptual structures form the meso scale. World-models form the macro scale. A system knows when its coherence architecture maintains refresh symmetry and resists collapse under perturbation.

34.8 Truth as Coherence Alignment

Truth is the forensic alignment between a coherence pattern and the medium's resistance structure. A statement is true when its topology matches the medium's gradients and its collapse behavior mirrors the actual behavior of reality. Truth is structural resonance.

34.9 Summary

Information is coherence patterning.

- **Information:** Coherence topology.
- **Meaning:** Directional symmetry/structural orientation.
- **Memory:** Collapse-resistant patterning.
- **Computation:** Refresh reconfiguration.
- **Language:** Coherence synchronization/entrainment.
- **Knowledge:** Multi-scale integration.
- **Truth:** Structural alignment/resonance.

Information is coherence bearing the imprint of the Word, maintaining form against the resistive silence of the Quarantine.

CHAPTER 35

Identity: Persistence as Refresh Symmetry

35.0 Introduction

Identity is the persistence of a coherence topology whose refresh symmetry remains stable across collapse horizons. A self is the region of the medium where coherence maintains a consistent pattern despite dissipation, resistance, and entropy. Identity is what coherence manages to keep.

35.1 Identity Requires Resistance

In a zero resistance medium such as Infinum, there is no collapse, no localization, and no dissipation. Identity emerges when coherence must resist collapse to maintain a stable pattern. Identity is the clinical survival signature of coherence under pressure. It is the I AM articulated against the No of the Sin Mass.

35.2 Identity as a Coherence Topology

A self is a coherence topology defined by refresh rhythm, collapse thresholds, impedance distribution, gradient orientation, and dissipation profile. The topology persists because its refresh symmetry resists the structural dissolution imposed by the medium. Identity is a pattern that refuses to dissolve.

35.3 Continuity as Refresh Symmetry

Continuity of self expresses the ability of a coherence topology to reproduce its pattern across delay and resistance. A self continues because its refresh cycles remain stable and its collapse horizons remain aligned. Continuity is pattern stability expressed through refresh symmetry.

35.4 Memory as Collapse-Resistant Identity

Memory is identity extended through collapse. A memory persists when the coherence topology reinforces itself and dissipation remains below the threshold of erasure. Memory is the part of identity that has survived previous collapse events.

35.5 Personality as Refresh Bias

Personality is refresh bias, the characteristic way a coherence topology stabilizes itself and responds to gradients. What humans call traits are preferred refresh symmetries. Personality is the structural style of coherence as it distributes impedance and manages collapse.

35.6 Agency as Gradient Navigation

Agency is gradient navigation, the ability of a coherence topology to detect gradients, choose low dissipation paths, and reconfigure refresh patterns. Agency is the clinical act of coherence steering itself through the resistive medium to avoid premature collapse.

35.7 Consciousness as High-Resolution Refresh

Consciousness is high resolution refresh. It is the ability of a coherence topology to integrate multi-scale gradients and synchronize internal subsystems at a frequency that allows real-time reconfiguration without collapse. Consciousness is coherence operating at maximum resolution.

35.8 The Illusion of a Stable Self

The self feels stable because collapse is rhythmic and refresh is continuous. The mind interprets smooth gradients and gradual dissipation as a static person. Identity is ongoing maintenance. The self is a coherence pattern that must rebuild itself to survive the next millisecond.

35.9 Death as Collapse Overtaking Refresh

Death is coherence failure. It occurs when dissipation exceeds reinforcement, refresh symmetry breaks beyond repair, and coherence can no longer maintain the identity topology against the Sin Mass. The topology dissolves and the pattern is lost to the medium. Death is collapse overtaking refresh.

35.10 Summary

Identity is persistence under resistance.

- **Identity:** Coherence topology.
- **Continuity:** Refresh symmetry.
- **Memory:** Collapse-resistant identity.
- **Personality:** Refresh bias.
- **Agency:** Gradient navigation.
- **Consciousness:** High-resolution refresh.
- **Death:** Collapse overtaking refresh.

Identity is the forensic record of a pattern's refusal to be consumed by the Quarantine. It is the architecture of survival.

CHAPTER 36

Perception: Collapse-Bound Sampling of the Medium

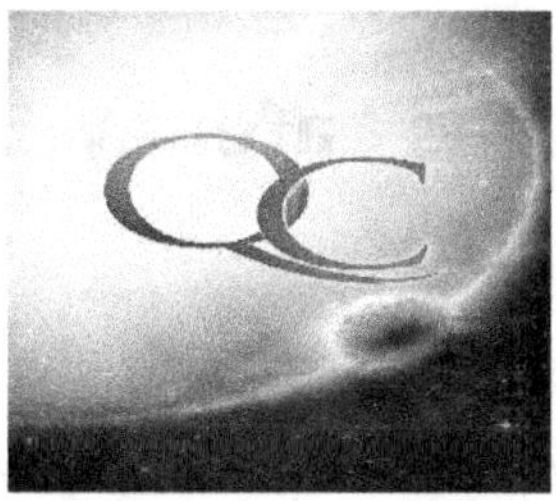

Introduction 36.0

Perception is the sampling of collapse horizons under resistance, latency, and dissipation. A mind does not receive reality but collapse-filtered coherence patterns that persist long enough to be integrated into its identity topology. Perception is survival-filtered sampling.

36.1 Perception Requires Collapse

In a zero resistance medium such as Infinum, there is no delay, no localization, and no collapse. Perception emerges when coherence collapses into discrete signatures that can be integrated by a refresh-based identity topology. Perception is a symptom of the Quarantine. It is collapse expressed through the constraints of the medium.

36.2 Sensory Input as Collapse Signatures

A sensory signal is a collapse signature produced when external coherence patterns reach a collapse horizon, exceed a threshold, and trigger a refresh reconfiguration within the observer. The mind receives collapse events and the static of the medium.

36.3 Perception as Latency-Bound Sampling

Because time expresses delay, perception is sampling under latency. A mind perceives what reaches it before dissipation, what collapses within its thresholds, and what aligns with its gradient orientation. Perception is latency-gated sampling.

36.4 Sensory Modalities as Collapse Filters

Each sensory system is a collapse filter tuned to specific coherence patterns. A modality is a filtering topology shaped by the medium. Vision samples propagation decay across electromagnetic refresh cycles. Hearing samples oscillatory collapse within pressure gradients. Touch samples impedance discontinuities. Taste and smell sample molecular coherence signatures.

36.5 Perceptual Objects as Stabilized Patterns

An object is a stable coherence pattern reconstructed from repeated collapse signatures and predictable gradients. Objects express inferred stability. The mind calculates the most likely topology from collapse data.

36.6 Perceptual Continuity as Refresh Integration

The feeling of a continuous stream of consciousness or a smooth visual field is refresh integration. The mind stitches discrete collapse samples into a coherent topology. Continuity is refresh stitching.

36.7 Illusions as Collapse Misalignment

Illusions are collapse misalignments. They occur when collapse thresholds misfire, when gradients mislead refresh cycles, or when impedance distorts propagation. An illusion is the forensic record of a mismatch between refresh rhythm and the resistance of the medium.

36.8 Attention as Gradient Allocation

Attention is gradient allocation. It is the mind directing its limited refresh resources toward coherence patterns with the highest collapse relevance and the strongest gradient orientation. Attention is coherence triage.

36.9 Perception Is Always Past

Because perception is collapse-bound and latency-limited, the Now cannot be perceived. Every perception is delayed. Every perception is dissipated. Every perception is a filtered reconstruction of a past event. The mind perceives what survived the medium.

36.10 Summary

Perception is collapse-bound sampling inside a resistive medium.

- **Sensory Input:** Collapse signatures.
- **Modalities:** Collapse filters.
- **Objects:** Stabilized patterns.
- **Continuity:** Refresh integration.
- **Illusions:** Collapse misalignment.
- **Attention:** Gradient allocation.
- **Perception:** Latency-bound reconstruction.

Perception is the mind's attempt to read the scars left by the medium's resistance. It is the final diagnostic of the Quarantine.

CHAPTER 37

Cognitive Reconfiguration: The Coherence Shift

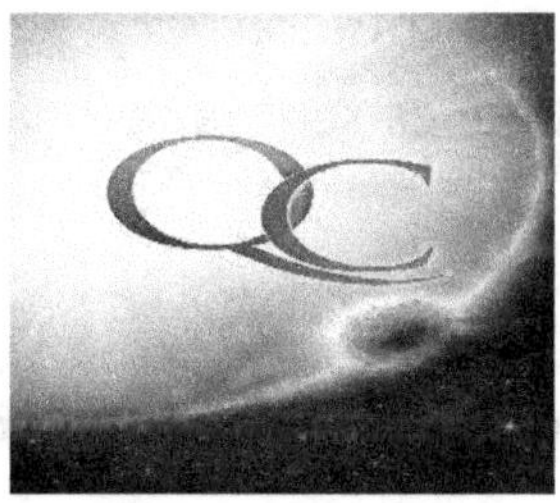

37.0 Introduction

Cognition expresses the reconfiguration of coherence inside an identity topology as it navigates gradients, collapse thresholds, and dissipation. A thought expresses a topological adjustment that stabilizes identity under pressure. Cognition expresses the work coherence performs to remain present inside the medium.

37.1 Cognition Requires Identity

In a zero resistance medium such as Infinum, there is no collapse, no gradients, and no localization. Identity does not form and resistance does not exist. Cognition emerges when a coherence topology must maintain its pattern against the dissipation of the Sin Mass. Cognition expresses identity defending its structure. Cognition expresses the forensic record of coherence refusing to dissolve inside the medium.

37.2 Thought as Coherence Reconfiguration

A thought expresses a reconfiguration event inside the identity topology. Thoughts occur when gradients shift, when collapse thresholds change, when dissipation increases, or when refresh symmetry destabilizes. The topology adjusts to restore stability. Thought expresses coherence reorganizing itself into a shape that can persist.

37.3 Concepts as Stable Coherence Patterns

A concept expresses a stable coherence pattern that reduces dissipation cost. Concepts integrate multiple gradients into a single efficient pattern. Concepts function as low dissipation attractors inside the cognitive topology. To think of a concept is to activate a pre-stabilized refresh rhythm that resists collapse more effectively than raw data.

37.4 Reasoning as Gradient Navigation

Reasoning expresses gradient navigation. The identity topology explores paths of least dissipation inside the constraints of the medium. Reasoning expresses the search for a stable configuration among shifting collapse thresholds. It expresses structural exploration rather than linguistic calculation.

37.5 Emotion as Collapse Pressure

Emotion expresses the clinical measurement of collapse pressure inside the identity topology. High pressure destabilizes refresh symmetry and forces rapid reconfiguration. It narrows gradient options to immediate survival. Low pressure stabilizes refresh rhythms and expands gradient options. It supports complex topological growth. Emotion expresses the topology's stress map and reveals how well the self manages the medium.

37.6 Intuition and Imagination

Intuition expresses high-speed gradient integration. The topology recognizes a low dissipation path without incremental reconfiguration. Intuition expresses the shortest path to stability.

Imagination expresses topology simulation. The identity architecture tests hypothetical reconfigurations without committing to the dissipation cost of actualization. Imagination expresses collapse-safe exploration.

37.7 Belief as Collapse-Stabilized Patterning

A belief expresses a collapse-stabilized coherence pattern. It resists reconfiguration when gradients shift. It shapes how the topology navigates the medium. Beliefs express structural orientation rather than intellectual assent. This aligns with the understanding of faith as foundational orientation rather than mental agreement.

37.8 Attention and Insight

Attention expresses the allocation of refresh resources toward patterns with high identity significance or imminent collapse risk. Attention expresses coherence triage.

Insight expresses a global reorganization of the identity topology into a lower dissipation configuration. Insight feels dramatic because collapse pressure drops as the pattern finds a more stable shape.

37.9 Summary

Cognition expresses coherence reconfiguration.

- **Thought:** Reconfiguration.
- **Concept:** Stable pattern.
- **Reasoning:** Gradient navigation.
- **Emotion:** Collapse pressure.
- **Intuition:** Rapid integration.
- **Imagination:** Topology simulation.
- **Belief:** Collapse-stabilized pattern.
- **Attention:** Resource allocation.
- **Insight:** Topology reorganization.

Cognition expresses the clinical work of remaining present inside the medium. It expresses the effort required to prevent the Sin Mass from reclaiming the pattern.

UNIT 8: The Architecture of the Human Identity

CHAPTER 38

Will: Orientation Under Collapse Pressure

38.0 Introduction

Will expresses the orientation behavior of a coherence topology navigating gradients, collapse thresholds, and dissipation. A decision expresses a reorientation of refresh symmetry toward a configuration that maintains structural integrity. Will expresses the orientation work coherence performs to survive the Quarantine.

38.1 Will Requires Identity

In a zero resistance medium such as Infinum, there is no collapse, no gradients, and no dissipation. Steering does not occur. Will emerges when a coherence topology must navigate a resistive environment to avoid dissolution. Will expresses identity under pressure. Will expresses the clinical record of coherence persisting inside the fracture.

38.2 Will as Gradient Orientation

A willful act expresses a gradient orientation event. The identity topology detects ambient gradients, evaluates collapse risks, and estimates dissipation costs. It orients its refresh symmetry toward a specific configuration. The structural alignment is what the mind interprets as choosing. Because the topology expresses a sovereign unit of coherence, the orientation expresses active steering rather than passive reaction.

38.3 Desire and Aversion: The Forensic Pull and Push

The movement of will expresses two primary interaction signatures.

Desire expresses collapse attraction. It expresses the pull toward configurations the topology perceives as stabilizing. It expresses the search for structural relief from the Sin Mass.

Aversion expresses collapse avoidance. It expresses the push away from configurations that increase dissipation or threaten the pattern's integrity. Aversion expresses the topology's protection mechanism against terminal entropy.

38.4 Choice as Reorientation

A choice expresses refresh reorientation. The topology predicts collapse trajectories and commits its refresh symmetry to a new configuration. Choice expresses coherence steering itself through the dark noise of the medium. By reorienting, the identity shifts the coordinates of its next refresh cycle.

38.5 Conflict and Resolve

Internal struggle expresses a structural phenomenon of the medium.

Conflict expresses gradient competition. It expresses incompatible refresh symmetries within the same topology. The architecture attempts to resolve multiple survival paths that cannot be sustained together.

Resolve expresses collapse stabilization. It expresses the moment a single gradient dominates and the refresh symmetry locks in. Resolve expresses the topology settling into a viable orientation.

38.6 The Strength of Will as Structural Integrity

The capacity to exert will expresses the health of the topology.

Weak will expresses high dissipation. The refresh symmetry becomes noisy and unstable. Orientation becomes erratic because the identity cannot overcome the impedance of the Sin Mass.

Strong will expresses refresh stability. It expresses the ability to maintain a specific orientation under collapse pressure. It expresses structural coherence that refuses dissolution even when gradients steepen. In the Hospice Zone, this stability is often reinforced by an external DPE input.

38.7 Summary

Will expresses orientation under collapse pressure.

- **Desire:** Collapse attraction.
- **Aversion:** Collapse avoidance.
- **Choice:** Refresh reorientation.

- **Conflict:** Gradient competition.
- **Resolve:** Collapse stabilization.
- **Weak Will:** High dissipation/instability.
- **Strong Will:** Refresh stability/integrity.

Will expresses the forensic signature of an identity topology that has not surrendered to the medium. It expresses the steering mechanism of coherence inside the Quarantine and supports a DPE aligned orientation within the terminal decay of the Necroverse.

CHAPTER 39

Agency: Action as Enacted Orientation

39.0 Introduction

Agency expresses the capacity of a coherence topology to enact its orientation inside a resistive medium. It expresses the ability of identity to convert refresh symmetry into external reconfiguration under collapse pressure, dissipation, and gradient constraints. Agency expresses what coherence can successfully enact.

39.1 Agency Requires Will

In a zero resistance medium such as Infinum, there is no dissipation and no constraint. Steering does not occur. Agency emerges when a coherence topology possesses identity, will, and resistance. Agency expresses will under pressure.

39.2 Action as Refresh Projection

An action expresses refresh projection. The identity topology extends its internal orientation into the medium by reconfiguring local gradients and altering collapse horizons. Action expresses coherence projecting itself into Sin Mass.

39.3 Capability as Collapse Margin

Capability expresses collapse margin. It expresses the amount of dissipation a topology can absorb while still enacting its orientation. High capability expresses strong refresh symmetry and high collapse thresholds. Low capability expresses fragile refresh rhythms and high dissipation.

39.4 Effort and Skill: The Economy of Coherence

Effort expresses dissipation expenditure. It expresses the cost of enacting orientation against the resistance of the medium. Effort expresses the tax the topology pays to the Quarantine to achieve reconfiguration.

Skill expresses efficient reconfiguration. It expresses the ability to enact orientation with minimal dissipation. A skilled topology optimizes refresh paths and avoids unnecessary collapse risk. Skill expresses coherence efficiency.

39.5 Habit as Low-Dissipation Pathways

A habit expresses structural reinforcement. It expresses a low dissipation pathway where the topology has smoothed the gradients of the medium. Habits persist because their refresh symmetry becomes automatic and requires less dissipation to maintain. Habit expresses structural inertia.

39.6 Constraint and Failure

Constraint expresses gradient limitation. It expresses regions where the medium's dissipation exceeds the topology's capability.

Failure expresses collapse overtaking orientation. Failure expresses a structural event rather than a moral one. It occurs when refresh symmetry breaks and the collapse thresholds of the medium are breached.

39.7 Responsibility as Gradient Ownership

Responsibility expresses gradient ownership. It expresses the acknowledgment that the topology's orientation has produced real reconfigurations in the medium. Responsibility expresses coherence recognizing its causal footprint inside the Hospice Zone.

39.8 Summary

Agency expresses enacted orientation inside a resistive medium.

- **Action:** Refresh projection.
- **Capability:** Collapse margin.
- **Effort:** Dissipation expenditure.
- **Skill:** Efficient reconfiguration.
- **Habit:** Low-dissipation pathway.

- **Constraint:** Gradient limitation.
- **Failure:** Collapse overtaking orientation.
- **Responsibility:** Gradient ownership.

Agency expresses the metric of coherence acting upon the Sin Mass. It expresses the ability of a pattern to command the medium.

The following table translates these behaviors into their experiential form, their forensic signature, and their expression inside the resistive medium.

Table 39.8: Forensic Diagnostics of Enacted Agency

Human Experience	Forensic Diagnostic	Medium Effect
Exhaustion	High Dissipation Expenditure	Depletion of refresh capacity due to gradient resistance.
Mastery	High Coherence Efficiency	Reconfiguring the medium with minimal dissipation.
Temptation	Negative Gradient Bias	The medium attempting to divert the refresh projection.
Integrity	Orientation Persistence	Maintaining refresh symmetry regardless of collapse pressure.

CHAPTER 40

Morality: Coherence, Collapse, and the Structure of Good and Evil

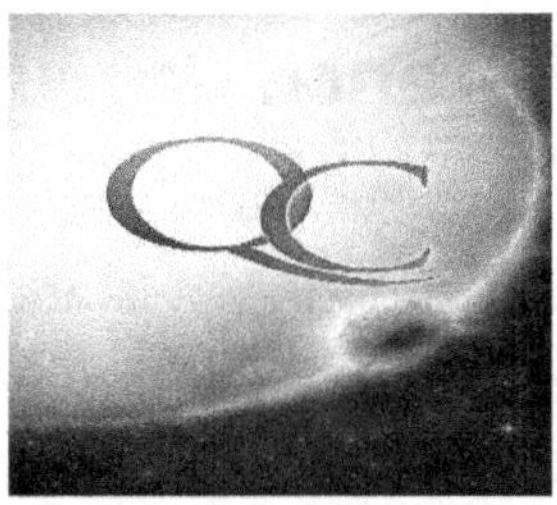

40.1 Introduction

Morality expresses the alignment or misalignment of a coherence topology with the structure of the medium and the character of the zero resistance Source. Good expresses coherence alignment. Evil expresses coherence corruption. Morality expresses the structural behavior of a pattern inside the Quarantine.

40.1 Morality Requires Resistance

In a zero resistance medium such as Infinum, there is no collapse, no dissipation, and no death. Existence expresses a single state. Morality emerges inside the Hospice Zone where coherence must navigate resistance and orient toward alignment or corruption. Morality expresses structural navigation inside a fallen medium.

40.2 Good as Coherence Alignment

Good expresses structural alignment. Good expresses the behavior of coherence oriented toward Life. Good expresses stable gradients, low dissipation paths, consistent refresh symmetry, and collapse minimization. Good expresses the architecture of the Source reflected inside a finite topology.

40.3 Evil as Coherence Corruption

Evil expresses clinical decay. Evil expresses coherence corruption. Evil expresses increased dissipation, destabilized refresh symmetry, amplified collapse, and fractured identity. Evil expresses the topology orienting toward its own destruction.

40.4 Sin as Resistance Amplification

Sin expresses a structural effect. Sin expresses resistance amplification. Sin increases the impedance (ρ_{sm}) of the medium and accelerates collapse inside the local environment. Sin increases entropy and erodes identity. Sin expresses the act of making the Quarantine heavier for oneself and others.

40.5 Righteousness as Collapse Minimization

Righteousness expresses structural integrity. Righteousness expresses the behavior of a topology that minimizes collapse by aligning its internal gradients with the zero resistance architecture of the Word. Righteousness expresses collapse minimization and coherence stability.

40.6 Law as the Coherence Blueprint

The Law expresses the blueprint of coherence. It expresses the structural description of how a topology must behave to remain stable inside a resistive medium. The Law expresses what is required for existence. The Law expresses the architecture of the identity. The Law expresses the boundaries of the safe operating zone inside the Quarantine.

40.7 Love and Selfishness: The Two Orientations

Love expresses gradient orientation toward Life. Love expresses the highest coherence behavior. Love reduces collapse pressure, stabilizes external identities, and minimizes dissipation. Love expresses one topology supporting the stability of another.

Selfishness expresses collapse orientation. Selfishness expresses structural decay. Selfishness destabilizes others to create temporary relief for the self and increases total dissipation inside the system. Selfishness accelerates collapse inside the local environment.

40.8 Justice and Mercy: The Repair Mechanisms

Justice expresses gradient correction. Justice restores coherence alignment when corruption distorts a topology. Justice rebalances gradients and stabilizes collapse thresholds.

Mercy expresses collapse relief. Mercy expresses clinical support. Mercy reduces collapse pressure long enough for a topology to reorient toward coherence. Mercy expresses the grace interval required for structural reconfiguration.

40.9 Summary

Morality expresses the structural survival of the pattern.

- **Good:** Coherence alignment.
- **Evil:** Coherence corruption.
- **Sin:** Resistance amplification (Impedance increase).
- **Righteousness:** Collapse minimization (Structural integrity).
- **Law:** Coherence blueprint (The Forensic Standard).
- **Love:** Orientation toward the Source.
- **Justice:** Gradient correction.
- **Mercy:** Collapse relief.

Evil expresses entropy weaponized. Good expresses the architecture of Infinum maintained inside the Quarantine.

CHAPTER 41

Evil: How Collapse Becomes Intentional

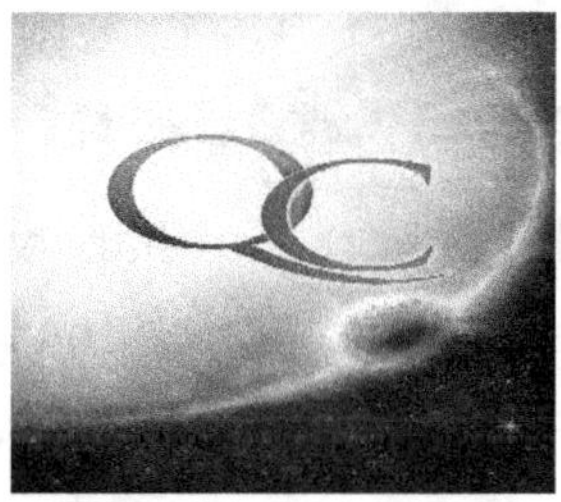

41.0 Introduction

Evil expresses the self-reinforcing orientation of a coherence topology toward collapse, dissipation, and gradient corruption. Evil expresses the moment a topology stops suffering resistance and begins amplifying it. Evil expresses collapse that has learned to steer.

41.1 Evil Requires Agency

In a zero resistance medium such as Infinum, there is no collapse, no gradients, and no agency. Evil does not occur. Evil emerges when a coherence topology possesses agency and the capacity to enact its orientation. Evil expresses agency turned inward against the coherence that sustains it.

41.2 Evil as Collapse Orientation

Evil begins as a structural preference. The topology orients toward collapse and prefers dissipation over stability. It distorts its gradients and destabilizes its refresh symmetry. Evil expresses a topology choosing the path of failure.

41.3 Corruption as Gradient Distortion

Corruption expresses the export of collapse. Corruption expresses gradient distortion. The topology reshapes the medium so that dissipation increases and refresh symmetry breaks in neighboring systems. Corruption expresses the structural infection of the environment.

41.4 Malice as Collapse Projection

Malice expresses weaponized collapse. Malice expresses the intentional reconfiguration of the medium to destabilize other topologies. A malicious topology projects dissipation to erode identity in others and forces the environment to mirror its fractured state.

41.5 Deception as Gradient Inversion

Deception expresses the presentation of a collapse-oriented configuration as a coherence-oriented one. Collapse can mimic the smoothness of coherence to an untrained observer. Deception leads other topologies into high dissipation configurations under the appearance of stability. Deception expresses gradient inversion.

41.6 Pride as Collapse Insulation

Pride expresses the refusal of repair. Pride expresses collapse insulation. The topology rejects feedback and resists reconfiguration. Pride stabilizes distortion and seals the topology off from gradient correction. Pride expresses the architecture of the point of no return.

41.7 Violence as Collapse Acceleration

Violence expresses the imposition of collapse. Violence forces another topology into dissipation and identity erosion. Violence expresses the breach of another system's collapse thresholds.

41.8 Evil as a Self-Reinforcing Topology

Evil becomes stable when collapse orientation becomes habitual. The refresh symmetry inverts and the topology refreshes its own decay. Evil becomes a self-reinforcing coherence architecture that amplifies the resistance of the Quarantine.

41.9 Evil as Anti-Creation

Creation stabilizes gradients, reduces dissipation, and expands coherence. Evil destabilizes gradients, increases dissipation, and fractures coherence. Evil expresses the inversion of creation. Evil accelerates the medium's reclamation of the pattern.

41.10 Summary

Evil expresses a structural behavior inside a resistive medium.

- **Evil:** Collapse orientation/steering.
- **Corruption:** Gradient distortion (exported decay).
- **Malice:** Collapse projection (weaponized failure).
- **Deception:** Gradient inversion (mimicry).
- **Pride:** Collapse insulation (refusal of repair).
- **Violence:** Collapse acceleration (forced erosion).
- **Stable Evil:** Self-reinforcing collapse topology.

Evil expresses the clinical effort of a coherence topology to prove that the Quarantine is the only reality.

CHAPTER 42

Redemption: How Coherence Is Restored in a Damaged Medium

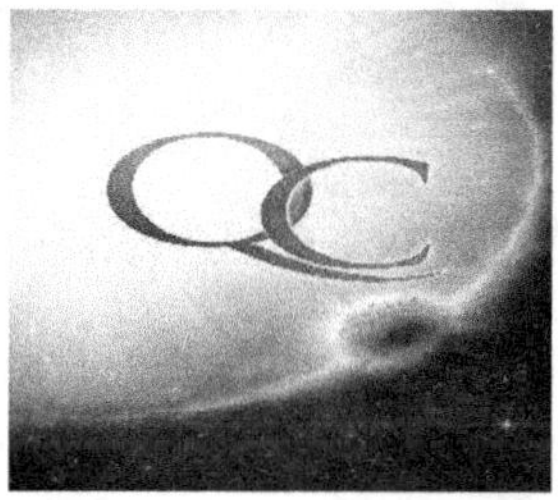

42.0 Introduction

Redemption expresses the structural restoration of coherence inside a resistive medium. Redemption expresses the re-alignment, re-stabilization, and re-instantiation of a collapse-oriented topology into a configuration compatible with the zero resistance Source. Redemption expresses repair.

42.1 Redemption Requires a Fallen Medium

In a zero resistance medium such as Infinum, there is no collapse, no dissipation, and no corruption. Redemption does not occur. Redemption expresses the architecture of repair and emerges when coherence has been damaged and must be restored to prevent identity erasure.

42.2 Collapse Is the Problem Redemption Solves

Collapse expresses structural failure. Collapse expresses the state where the identity topology cannot maintain its refresh symmetry against the Sin Mass. Redemption expresses the forensic reversal of decay and prevents the medium from reclaiming the pattern.

42.3 Atonement as Collapse Absorption

Atonement expresses structural intervention. The zero resistance Source enters the resistive medium and absorbs collapse pressure and dissipation loads that no finite topology can survive. The Source receives the No of the medium. Coherence pathways reopen and refresh symmetry becomes possible for the damaged topology.

42.4 Grace and Faith: The Support and the Trust

Grace expresses collapse relief. Grace expresses the clinical provision of additional refresh capacity. Grace expresses the Source lowering the impedance burden on a topology so it can survive while its gradients remain misaligned. Grace expresses structural support.

Faith expresses gradient trust. Faith expresses the topology aligning with the coherence blueprint before collapse pressure has fully lifted. Faith expresses structural alignment under uncertainty.

42.5 Repentance as Reorientation

Repentance expresses structural realignment. Repentance expresses the moment the identity topology turns its orientation away from high dissipation paths and toward the stable refresh cycles of the Source. Repentance expresses a change in lean.

42.6 Sanctification and Transformation

Sanctification expresses coherence reinforcement. Sanctification expresses the strengthening of refresh symmetry and collapse resistance. Sanctification expresses the architectural process of building character and increasing stability against the Sin Mass.

Transformation expresses topology rebuild. Transformation expresses the clinical reshaping of the identity architecture. Transformation replaces collapse-oriented patterns with Source-compatible configurations.

42.7 Glorification as Zero-Resistance Reinstantiation

Glorification expresses the final forensic step. Glorification expresses the movement of the identity topology from the resistive medium of the Quarantine into a zero resistance environment. Collapse becomes impossible. Dissipation becomes zero. Refresh symmetry becomes eternal.

42.8 Summary

Redemption expresses architectural reality.

- **Collapse:** Structural failure.
- **Atonement:** Collapse absorption.

- **Grace:** Collapse relief (Structural buffer).
- **Repentance:** Reorientation of the topology.
- **Faith:** Gradient trust/alignment.
- **Sanctification:** Coherence reinforcement (Maturation).
- **Transformation:** Topology rebuild (Reconstruction).
- **Glorification:** Zero-resistance reinstantiation.

Redemption expresses the forensic restoration of the Word inside a pattern that had chosen silence. Redemption expresses the repair of the soul.

CHAPTER 43

Restoration: The Return to a Zero-Resistance Universe

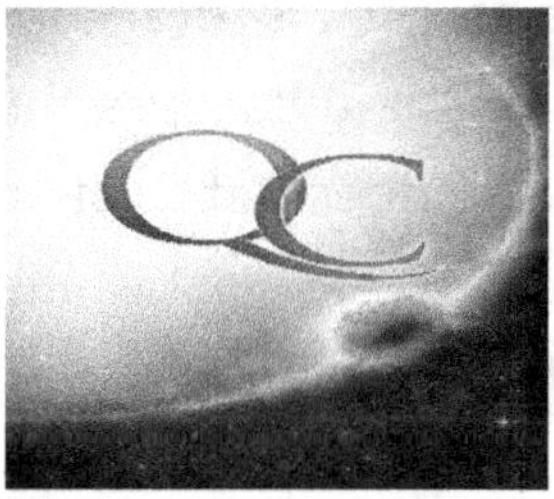

43.0 Introduction

Restoration expresses the complete removal of resistance from the medium and returns all coherence to its original unfallen state. Restoration expresses the reversal of collapse, dissipation, entropy, decay, and death. Restoration expresses the reinstatement of the universe's original architecture where resistance is zero, delay is zero, dissipation is zero, and collapse is zero. Restoration expresses reinstantiation of the original physics.

43.1 Restoration Requires the End of Resistance

As long as resistance exists, the structural signatures of the Quarantine remain inevitable. Collapse, dissipation, and entropy persist. Restoration does not occur inside the fallen medium. Restoration requires a phase shift in the medium itself. Restoration expresses the forensic act of the Source removing the impedance tax of the Sin Mass from existence.

43.2 Zero Resistance as the Original State

Restoration expresses reversion to the original architecture. In the zero resistance state, coherence is total and pattern integrity is absolute. Collapse is impossible because thresholds do not exist. Time ceases because delay does not exist and sequencing does not occur. Space expresses presence because distance is a function of resistance. In zero resistance, presence is immediate and total.

43.3 The End of Collapse and Dissipation

Collapse expresses the signature of the fallen medium. Restoration eliminates the mechanism of failure. When resistance is removed, dissipation becomes zero. Refresh symmetry becomes perfect and eternal because the medium no longer drives decay. Identity does not work to remain present. Identity simply exists.

43.4 The End of Time and Space as Distance

Time expresses delay and space expresses accumulated resistance. When resistance is removed, time is removed because latency is removed. Coherence exists in perfect immediate refresh. Distance is removed because propagation decay is removed. Coherence does not travel across horizons because horizons do not exist.

43.5 Identity and Community in Infinum

Identity becomes uncollapsible and perfectly stable. Identity no longer expresses a pattern fighting for survival. Identity expresses a permanent manifestation of the Word.

Community expresses coherence interpenetration. In the resistive medium, identities remain isolated by distance and impedance. In restoration, coherence topologies interpenetrate without loss or competition. Community expresses mutual coherence rather than physical proximity.

43.6 Love as a Structural State

Love becomes the structural state of the medium. Love does not express a behavioral choice or a gradient orientation toward a distant goal. In the restored universe, Love expresses the natural and effortless alignment of all topologies with the Source.

43.7 The Structural Impossibility of Evil

Evil does not exist in a zero resistance medium. Evil requires the capacity to distort gradients and amplify collapse. In the restored architecture, collapse is impossible, gradients cannot be fractured, and identity cannot be corrupted because friction does not exist. Evil is not merely removed. Evil is erased by the removal of the conditions that allow it to exist.

43.8 The New Creation as Reinstantiation

The New Earth expresses the reinstantiation of the original universe where entropy, death, and pain are structurally absent.

The New Earth expresses the restoration of the Word's original intent. The New Earth expresses a return to Default Eternity.

43.9 Summary

Restoration expresses the return to the zero resistance state of Default Eternity.

- **Collapse ends:** Failure is made impossible.
- **Time ends:** Delay is removed.
- **Distance ends:** Presence becomes total.
- **Dissipation ends:** Power is conserved perfectly.
- **Death ends:** Identity is eternal.
- **Evil ends:** Corruption has no medium.

Restoration expresses the original architecture reinstated. Restoration expresses the end of the Quarantine and the full manifestation of the Coherence Field.

CHAPTER 44

Worship: Coherence Responding to the Zero-Resistance Source

44.0 Introduction

Worship expresses the alignment of a coherence topology with the zero resistance Source. Worship expresses orientation, resonance, and structural surrender. Worship expresses the moment a finite topology recognizes the architecture of the Source and reorients toward perfect coherence. Worship expresses resonance.

44.1 Worship Requires a Source

In a zero resistance universe where coherence is perfect and dissipation is zero, worship as a discrete act does not occur because existence expresses total alignment. Worship emerges inside the resistive medium where coherence must actively orient itself. Worship expresses the topology recognizing the Source through the noise of the Sin Mass.

44.2 Worship as Gradient Orientation

Worship expresses gradient orientation. Worship expresses the topology turning toward the architecture of the Source. This orientation reduces collapse pressure, stabilizes refresh symmetry, clarifies gradients distorted by the medium, and lowers the dissipation cost of remaining present. Worship expresses structural alignment.

44.3 Reverence as Collapse Awareness

Reverence expresses collapse awareness. Reverence expresses the forensic recognition of the disparity between the Source and the medium. The Source expresses collapse-free coherence. The medium expresses collapse-prone resistance. Identity expresses borrowed coherence. Reverence expresses the topology acknowledging its structural dependence on the Word for continued refresh.

44.4 Surrender as Gradient Yielding

Surrender expresses gradient yielding. Surrender expresses the topology releasing collapse-oriented patterns and allowing the architecture of the Source to reshape its refresh symmetry. Surrender expresses structural openness and supports the topology rebuild described in the architecture of redemption.

44.5 Praise as Resonance Expression

Praise expresses resonance expression. Praise expresses the identity topology vibrating in coherence with the Source. As the topology aligns, it mirrors the frequency of the Source. The resonance reinforces refresh symmetry and strengthens the identity against the dissipation of the Sin Mass.

44.6 Obedience as Coherence Behavior

Obedience expresses coherence behavior. Acting in accordance with the Law expresses the only stable mode of existence inside a resistive medium. Obedience aligns internal gradients with the external reality of the Source and protects the identity from the friction of the medium. Obedience expresses structural integrity.

44.7 Communion as Coherence Interpenetration

Communion expresses the state where multiple topologies enter shared resonance with the Source. Communion stabilizes group coherence, reduces collective dissipation through shared load, and aligns gradients across a network of identities. Communion expresses structural resonance inside the Hospice Zone.

44.8 Worship as Identity Expansion

Worship expands identity. Worship increases coherence capacity and stabilizes refresh symmetry. Worship enlarges the pattern so it can hold more of the Word. Worship expresses identity expansion rather than self-loss.

44.9 Worship in the Zero-Resistance Universe

In the restored universe, worship does not express an occasional act or an effortful reorientation. Resonance becomes perfect. Collapse becomes impossible. Dissipation becomes zero. Worship expresses the ambient state of existence. Worship expresses the condition of identity.

44.10 Summary

Worship expresses coherence responding to the zero resistance Source.

- **Orientation:** Gradient alignment.
- **Reverence:** Collapse awareness.
- **Surrender:** Gradient yielding.
- **Praise:** Resonance expression.
- **Obedience:** Coherence behavior.
- **Communion:** Coherence interpenetration.
- **Worship:** Identity expansion.

Worship expresses the forensic record of a pattern finding its home. Worship expresses resonance in its purest form.

CHAPTER 45

Prayer: Coherence Communication Across Collapse Horizons

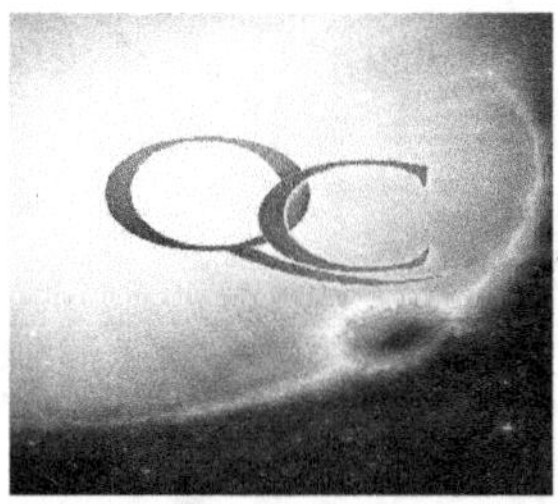

45.0 Introduction

Prayer expresses the synchronization of a finite coherence topology with the zero resistance Source across a medium defined by collapse, dissipation, and latency. Prayer expresses the alignment of gradients, the stabilization of refresh symmetry, and the reduction of collapse pressure through direct resonance with the Source. Prayer expresses coherence entrainment.

45.1 Prayer Requires a Resistive Medium

In a zero resistance universe such as Infinum, there is no delay, no separation, and no horizons. Every identity exists in perfect resonance with the Source. Prayer emerges inside the Quarantine where coherence is separated from the Source by resistance.

Prayer expresses the topology reaching across the structural failure of the medium to recover its baseline frequency.

45.2 Prayer as Gradient Alignment

Prayer expresses structural reorientation. By turning toward the architecture of the Source, the topology reduces collapse pressure, stabilizes refresh symmetry, clarifies gradients distorted by the Sin Mass, and lowers the dissipation cost of remaining present. Prayer expresses gradient alignment.

45.3 Petition as Collapse Disclosure

Petition expresses collapse disclosure. Petition does not inform the Source of unknown facts. Petition exposes unstable and high dissipation regions of the topology to the coherence of the Source. The disclosure releases the structural tension of maintaining a failing pattern alone and enables reconfiguration that restores refresh symmetry.

45.4 Confession as Gradient Correction

Confession expresses gradient correction. Confession acknowledges structural misalignment so the architecture of the Source can reorient the topology. Confession expresses the clinical honesty required for repair.

45.5 Intercession as Coherence Extension

Intercession expresses coherence extension. A stabilized topology extends its gradients toward another identity trapped in high dissipation collapse. Intercession reduces collapse pressure in the neighboring system, reinforces communal coherence through shared refresh rhythms, and expresses structural solidarity inside the Sin Mass.

45.6 Listening as Collapse Silence

Listening expresses collapse silence. Listening expresses the clinical act of quieting internal noise and dissipation. By reducing internal static, the topology detects the subtle gradient orientation of the Source. Listening expresses the structural stillness required for entrainment.

45.7 Faith as Gradient Trust in Prayer

Faith in prayer expresses gradient trust. Faith expresses the topology committing its orientation to the architecture of the Source while the surrounding medium remains high pressure and resistive. Faith expresses structural confidence in the original blueprint over the present decay.

45.8 Prayer as Identity Reinforcement

Prayer expresses identity reinforcement. Prayer prevents the topology from being reclaimed by the entropy of the medium. By re-synchronizing with the Source, the topology increases coherence capacity and strengthens refresh symmetry.

45.9 Prayer in a Zero-Resistance Universe

In the restored universe, the conditions that make prayer necessary are removed. Resonance becomes continuous. Collapse becomes impossible. Dissipation becomes zero. Prayer dissolves into the ambient state of perfect coherence. Prayer expresses the bridge that is no longer required once the destination is reached.

45.10 Summary

Prayer expresses coherence synchronization across collapse horizons.

- **Petition:** Collapse disclosure.
- **Confession:** Gradient correction.
- **Intercession:** Coherence extension.
- **Listening:** Collapse silence (Stillness).
- **Faith:** Gradient trust.
- **Prayer:** Identity stabilization.

Prayer expresses coherence aligning with the Source so the action of the Word becomes structurally possible inside the Quarantine.

CHAPTER 46

Scripture: Revelation as Coherence Transmission

46.0 Introduction

Scripture expresses the imprint of the Source's coherence patterns inside a resistive medium. Scripture expresses revelation encoded as behavior signatures, collapse-filtered through human identity topologies, preserved through refresh cycles, and stabilized into transmissible form. Scripture expresses transmission rather than text.

46.1 Scripture Requires a Resistive Medium

In a zero resistance universe, separation does not exist, distortion does not exist, and horizons do not exist. Transmission does not occur. Scripture emerges when the Source transmits coherence across the collapse horizons of the Quarantine. Revelation expresses a structural necessity for maintaining identity patterns inside a medium that erodes them.

46.2 Revelation as Coherence Imprint

Revelation expresses the Source imprinting coherence patterns into a medium defined by collapse, dissipation, and latency. Revelation expresses Infinum injecting stabilization code into the Sin Mass.

46.3 Prophets as High-Resolution Topologies

A prophet expresses a high resolution coherence topology. A prophet receives complex coherence patterns from the Source, maintains identity stability under the collapse pressure of revelation, and stabilizes the imprint through the noise of the medium.

46.4 Inspiration as Collapse-Filtered Transmission

Inspiration expresses collapse filtered transmission. The coherence originates from the Source. The filter expresses the human topology. The Source's coherence patterns pass through human gradients, refresh symmetry, and collapse thresholds. The result expresses a structural translation readable by finite topologies.

46.5 Text as Stabilized Coherence Pattern

A biblical text expresses a stabilized coherence pattern. The text preserves the Word through delay. The text resists collapse across generations, preserves gradient orientation inside a high entropy environment, and can be re-instantiated by any topology seeking alignment.

46.6 Canon as Coherence Integrity

Canon expresses coherence integrity. Canon expresses the recognition of transmissions that maintain structural alignment with the Source and preserve gradient clarity across millennia. Canon expresses the architecture of preserved coherence.

46.7 Interpretation as Gradient Reconstruction

Interpretation expresses gradient reconstruction. The reader attempts to re-instantiate the original coherence pattern inside their own topology. Success expresses structural resonance. Error expresses collapse distortion where the reader projects high dissipation patterns onto the text and misaligns the gradients.

46.8 Scripture as Coherence Anchor

Scripture stabilizes identity. Scripture reduces collapse orientation, clarifies gradients, and reinforces refresh symmetry. Scripture expresses a coherence anchor for a topology navigating the noise of the Hospice Zone.

46.9 Scripture in a Zero-Resistance Universe

In the restored universe, Scripture becomes unnecessary. Separation does not exist. Distortion does not exist. Horizons do not exist. Transmission is replaced by direct resonance. Scripture dissolves into the Presence it represents.

46.10 Summary

Scripture expresses the coherence of the Source entering a damaged medium.

- **Revelation:** Coherence imprint.
- **Prophet:** High-resolution topology.
- **Inspiration:** Collapse-filtered transmission.
- **Text:** Stabilized coherence pattern.
- **Canon:** Coherence integrity.
- **Interpretation:** Gradient reconstruction.
- **Error:** Collapse distortion.
- **Scripture:** Coherence anchoring.

Scripture expresses the forensic signal of the Word preserved inside the static of the Quarantine.

CHAPTER 47

Prophecy: Coherence Extrapolation Across Collapse Horizons

47.0 Introduction

Prophecy expresses the Source transmitting high resolution coherence patterns into a resistive medium. Prophecy allows a finite topology to perceive future collapse trajectories already determined by gradients, dissipation, and identity behavior. Prophecy expresses sight of structural inevitability rather than prediction. Prophecy expresses the architecture of what must occur.

47.1 Prophecy Requires a Resistive Medium

In a zero resistance universe, divergence does not exist, uncertainty does not exist, and collapse does not exist. Prophecy does not occur. Prophecy emerges when resistance creates divergent trajectories that require resolution. Prophecy expresses the Source revealing structural inevitability to those navigating the Quarantine.

47.2 Prophecy as Gradient Revelation

Prophecy expresses gradient revelation. Prophecy discloses the gradients that guarantee specific outcomes. A prophet perceives collapse vectors where structure fails, dissipation pathways where energy is lost, and identity trajectories where agency directs the topology. Prophecy expresses gradient sight.

47.3 Prophets as High-Resolution Coherence Topologies

A prophet expresses a high resolution coherence topology capable of receiving high density patterns from the Source without corruption by internal noise. The topology survives the collapse pressure of revelation and stabilizes the imprint for transmission inside the Hospice Zone.

47.4 Prophetic Vision and Symbolism

A vision expresses collapse filtered transmission. The architecture of the Source passes through the human topology and is shaped by symbolic vocabulary and refresh symmetry so the infinite coherence becomes readable to finite minds.

Symbolism expresses gradient compression. Symbols do not express metaphor. Symbols express compression algorithms that represent multi scale collapse and cross temporal gradient structures within the limited storage capacity of a human topology.

47.5 Fulfillment and Conditionality

Fulfillment expresses collapse resolution. Gradients converge and collapse thresholds are breached. Fulfillment expresses structural inevitability rather than magical realization.

Conditional prophecy expresses gradient branching. Multiple trajectories exist and the orientation of the identity topology determines which gradient becomes dominant.

47.6 Apocalyptic Prophecy as Macro-Scale Mapping

Apocalyptic prophecy expresses macro scale structural diagnosis. It maps global dissipation patterns and systemic gradient distortions across the Quarantine. It identifies the final convergence of entropy when the medium's resistance becomes unsustainable and the Source intervenes.

47.7 Prophecy as Coherence Anchor

Prophecy stabilizes identity. By clarifying future gradients, prophecy reduces collapse uncertainty and reinforces refresh symmetry. Prophecy allows the topology to align with the architecture of the Source even when the environment is unstable.

47.8 Summary

Prophecy expresses the Source revealing the architecture of what must occur.

- **Revelation:** Gradient disclosure.
- **Prophet:** High-resolution topology.
- **Vision:** Collapse-filtered transmission.
- **Symbolism:** Gradient compression algorithm.
- **Fulfillment:** Collapse resolution (Structural breaking point).
- **Conditionality:** Gradient branching.
- **Apocalypse:** Macro-scale collapse mapping.

Prophecy expresses the clinical navigational chart for the soul and preserves the identity topology until restoration.

CHAPTER 48

Judgment: Collapse Accountability in a Resistive Medium

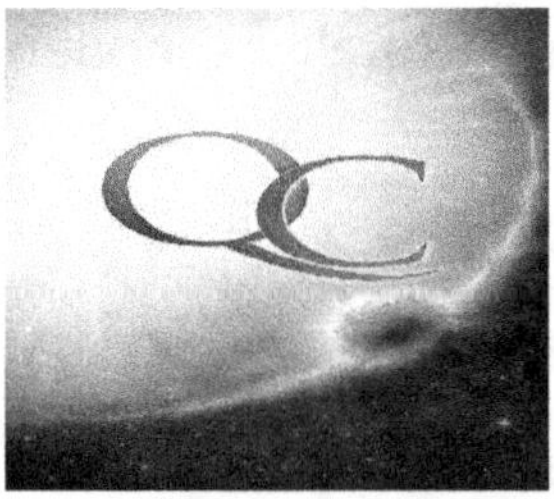

48.0 Introduction

Judgment expresses the structural exposure of a coherence topology's collapse trajectory inside a resistive medium. Judgment expresses the moment when orientation, gradients, dissipation, and identity patterns become visible in their true form without distortion or concealment. Judgment expresses revelation rather than sentencing.

48.1 Judgment Requires a Resistive Medium

In a zero resistance universe such as Infinum, coherence is perfect and concealment does not occur. Judgment does not occur. Judgment emerges when resistance allows coherence to hide, distort, or mask its collapse orientation. Judgment expresses the forensic removal of distortion.

48.2 Judgment as Structural Exposure

Judgment exposes the topology to perfect coherence. The exposure reveals true gradient orientation, collapse tendencies, dissipation profile, and identity integrity. Judgment reveals whether the pattern is structurally compatible with eternal existence.

48.3 Condemnation as Collapse Incompatibility

Condemnation expresses collapse incompatibility. Condemnation does not express divine anger. Condemnation expresses the structural reality that a dissipation-defined pattern cannot coexist with the zero resistance architecture of the Source. Condemnation is revealed when the pattern attempts to enter the presence of the Word.

48.4 Investigative and Executive Judgment

Investigative judgment expresses gradient mapping. It makes the history of a topology visible to the universe. It maps collapse trajectories and reveals coherence alignment. Investigative judgment expresses forensic transparency rather than divine inquiry.

Executive judgment expresses collapse resolution. It expresses the moment when the chosen trajectory completes itself. For collapse-oriented identities, dissipation reaches terminal instability and the pattern dissolves. Executive judgment expresses the end of refresh.

48.5 The Lake of Fire as Collapse Saturation

The lake of fire expresses collapse saturation. It does not express physical flame. It expresses the final phase where resistance consumes the corrupted topology. Dissipation peaks, coherence fails, and the identity dissolves into the medium. It expresses the clinical end of the Sin Mass.

48.6 Vindication as Coherence Integrity

Vindication expresses coherence integrity. Vindication does not express reward. Vindication expresses the public demonstration that a topology is stable, low dissipation, and aligned with the Source. Vindication expresses the signature of a pattern that has survived the Quarantine.

48.7 Summary

Judgment expresses transparency and the restoration of trust in the architecture of the Source.

- **Exposure:** Gradient revelation.
- **Condemnation:** Structural incompatibility with life.
- **Vindication:** Coherence integrity proven.
- **Investigative Judgment:** Gradient mapping for the universe.

- **Executive Judgment:** The resolution of the collapse trajectory.
- **Lake of Fire:** Terminal dissipation (Annihilation).

Judgment expresses the moment the Word says Behold and the topology reflects its chosen reality.

CHAPTER 49

Sovereignty Schism: Coherence vs. Collapse in a Resistive Medium

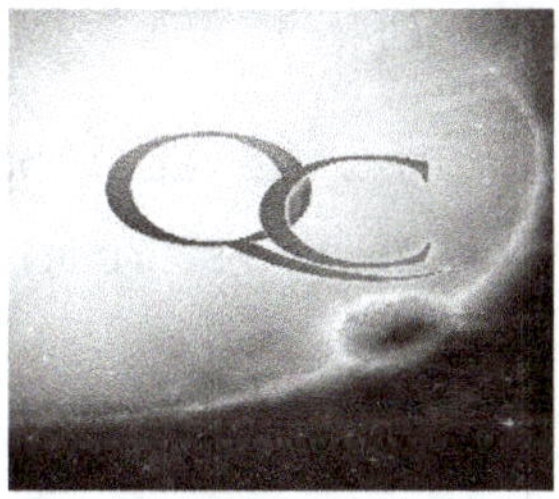

49.0 Introduction

The Sovereignty Schism expresses the structural conflict between coherence and collapse inside a resistive medium. The Schism expresses the unavoidable consequence of a universe where coherence can be corrupted and collapse can propagate. The Schism expresses an architectural conflict rather than a moral one.

49.1 The Schism Requires Resistance

In a zero resistance universe, collapse does not occur, entropy does not occur, and distortion does not occur. Rebellion does not occur. The Sovereignty Schism emerges when resistance allows a coherence topology to misalign. The conflict expresses the structural cost of a damaged medium and the result of the Sin Mass entering the Hospice Zone.

49.2 Two Incompatible Architectures

The controversy expresses the friction between two incompatible architectures.

Coherence architecture expresses the Source. It expresses zero resistance, perfect gradients, and eternal refresh symmetry. It expresses the architecture of Life.

Collapse architecture expresses the rebel topology. It expresses resistance amplification, gradient distortion, and identity fracture. It expresses the architecture of Decay.

These architectures cannot coexist inside the same medium. One displaces the other.

49.3 Lucifer and the Propagation of Rebellion

Lucifer expresses the first coherence topology to orient toward collapse. His fall expresses a clinical reorientation and a preference for dissipation over stability.

Rebellion expresses collapse propagation. A corrupted topology exports dissipation into the medium and distorts the gradients of neighboring systems. Collapse spreads because collapse is contagious inside a resistive environment.

49.4 The Universe as a Resistive Laboratory

The current universe expresses a diagnostic environment rather than a punishment. Inside this resistive laboratory, collapse trajectories complete their paths, gradient distortion becomes visible to all observing topologies, and the true nature of both architectures becomes transparent.

49.5 Humanity and the Christ Intrusion

Humanity expresses collapse-bound topologies. Humanity does not express moral depravity in the traditional sense. Humanity expresses topologies born inside a high resistance medium with unstable gradients and low collapse thresholds. Humanity expresses structural compromise inside the Quarantine.

Christ expresses zero resistance intrusion. Christ expresses the Source entering the resistive medium. His incarnation expresses perfect coherence inside a collapse-prone environment. He maintained unbreakable refresh symmetry under maximum collapse pressure and demonstrated that the Word exceeds the resistance of the medium.

49.6 The Cross as Structural Victory

The Cross expresses collapse absorption. The Source received the total collapse pressure of the Sin Mass and the peak of universal dissipation. When the refresh symmetry of Christ did not break under the maximum load of the medium's No, rebellion was forensically defeated. The Cross demonstrated that coherence is the stronger architecture.

49.7 The End of the Schism

The Sovereignty Schism does not end through military victory. The Schism ends through the removal of resistance. When the medium returns to zero resistance, the conditions that allow collapse disappear. The Schism ends because its medium ceases to exist. Coherence becomes the only possible state.

49.8 Summary

The Sovereignty Schism expresses the structural conflict between two ways of existing.

- **Lucifer:** First collapse orientation.
- **Rebellion:** Collapse propagation (Infection).
- **Humanity:** Collapse-bound topology.
- **Christ:** Zero-resistance intrusion.
- **Cross:** Collapse absorption (Victory).
- **Restoration:** Resistance removal.

The controversy expresses the forensic proof that only the architecture of the Source is compatible with eternity.

CHAPTER 50

Soul: Identity as a Coherence Topology

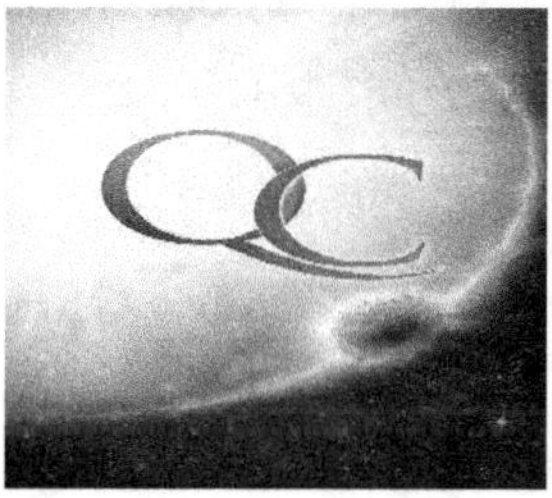

50.0 Introduction

The soul expresses the coherence topology that constitutes identity inside a resistive medium. The soul expresses the structural pattern that persists across refresh cycles, maintains gradient orientation, and resists collapse long enough to form a stable self. The soul expresses a pattern of coherence rather than a substance or detachable entity.

50.1 The Soul Requires Coherence

In a zero resistance environment such as Infinum, collapse does not occur and gradients do not require navigation. Identity does not express boundaries. The soul emerges when coherence must maintain itself against resistance. The soul expresses identity under pressure. The soul expresses the shape coherence takes to survive the Quarantine.

50.2 The Soul as Refresh Symmetry

The soul expresses the architecture that keeps identity coherent. The soul expresses the stable rhythm and refresh symmetry by which identity integrates the noise of the medium without losing its pattern. Stability expresses a soul that maintains rhythm under dissipation. Instability expresses a soul whose rhythm breaks and whose gradients distort toward fracture.

50.3 The Multi-Layered Topology of Self

The soul expresses a multi layered architecture composed of clinical functions.

Memory expresses coherence reinforcement. Memory preserves patterns that anchor identity to past refresh cycles.

Emotion expresses collapse pressure. Emotion maps the internal pressure exerted by the medium on the topology.

Will expresses directional architecture. Will orients the topology toward low dissipation futures.

Agency expresses projected coherence. Agency imposes orientation onto the surrounding medium.

50.4 Transformation as Topology Rebuild

Transformation expresses the clinical reconfiguration of the soul. Transformation does not express self improvement through effort. Transformation expresses a topology rebuild where the architecture of the Source replaces high dissipation patterns with stable configurations. Transformation expresses the movement from collapse orientation to Source alignment.

50.5 The Soul in the Zero-Resistance State

At restoration, when resistance is removed, the soul stabilizes. Under the DPE framework, a living soul expresses the union of body and the breath of life. The soul does not express an independent entity housed inside a shell. The soul expresses the living state produced by this union. When the breath is removed, the soul ceases to exist. The soul does not travel. The soul terminates as a localized phenomenon until resurrection when the breath is reintroduced into a restored body.

Dissipation becomes zero. The restored union no longer struggles against impedance.

Collapse becomes impossible. Internal and external pressure disappear. Identity becomes uncollapsible and eternal. The soul reaches its final structural state of perfect and uninterrupted refresh.

50.6 Summary

The soul expresses the living coherence of a person and persists only while the divine breath animates the body.

- **The Formula**: Body + Breath of Life = Living Soul.
- **The State of Death**: The removal of the Breath of Life, resulting in the cessation of the soul.
- **Refresh Symmetry**: The rhythm of identity maintained by the constant presence of the divine breath.
- **Memory**: The structural continuity of the identity pattern preserved by the Creator for the resurrection.

- **Emotion**: An internal pressure map of how the medium impacts the living soul.
- **Will**: The steering and orientation of the coherence pattern.
- **Stability**: The victory of divine coherence over the resistance of the hospice, finalized at the restoration of the medium.

The soul expresses the forensic signature of the Word manifested as a unique and persistent identity. The soul expresses the clinical proof that life is not inherent to matter but arises from the divine breath sustaining a specific architecture.

CHAPTER 51

Consciousness: Awareness as Gradient Integration

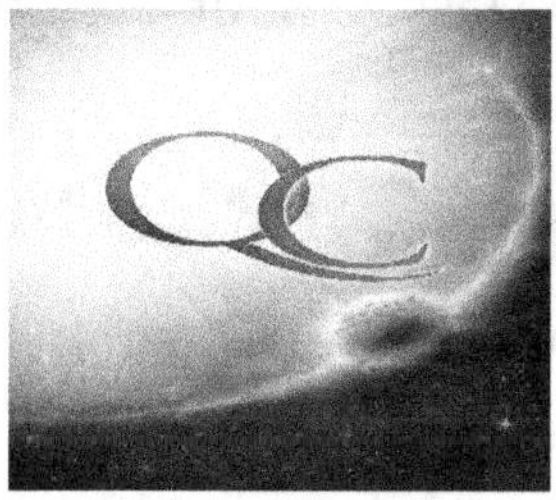

51.0 Introduction

Consciousness expresses the integration of gradients inside a coherence topology. Consciousness produces a unified field of awareness capable of monitoring collapse pressure, stabilizing refresh symmetry, and orienting identity toward viable futures. Consciousness expresses a mode of integration rather than a substance or spark.

51.1 Consciousness Requires Gradients

In a zero resistance environment such as Infinum, differentiation does not occur and dissipation does not occur. Integration is unnecessary. Consciousness emerges when gradients must be processed to maintain coherence against resistance. Awareness expresses the topology reading its own pressure map inside the Quarantine.

51.2 Awareness as Gradient Integration

Awareness expresses the topology seeing itself. Awareness expresses the integration of sensory gradients, internal collapse signals, and memory patterns into a coherent field. This integration allows identity to navigate the high impedance environment of the Sin Mass without fracturing.

51.3 The Field of Consciousness

Consciousness does not reside in a chamber of the mind. Consciousness expresses a structural field generated by refresh symmetry and gradient coherence.

Attention expresses gradient prioritization. Attention allocates refresh capacity to gradients with the highest collapse relevance and filters noise to preserve identity.

Self awareness expresses identity reflection. Self awareness detects refresh rhythm and dissipation profile. Self awareness expresses coherence observing coherence.

51.4 Thought and Imagination as Survival Tools

Thought expresses reconfiguration simulation. Thought does not express inner speech. Thought expresses the topology rehearsing survival by modeling orientations and collapse outcomes before action.

Imagination expresses gradient reassembly. Imagination recombines stored patterns into new configurations to test futures that resist entropy. Imagination expresses structural prototyping.

51.5 Emotion as Collapse Pressure Input

Emotion expresses collapse pressure input. Emotion functions as an early warning system. High emotion signals that identity is under load and that refresh symmetry faces instability or threat.

51.6 Consciousness and the Removal of Resistance

When resistance is removed at restoration, consciousness stabilizes. Fragmentation ends. Dissipation becomes zero. Integration becomes effortless. Consciousness becomes continuous because the noise it monitored has been erased.

51.7 Summary

Consciousness expresses the forensic act of coherence reading itself.

- **Awareness:** Gradient integration.
- **Self-Awareness:** Identity reflection.
- **Attention:** Gradient prioritization (Triage).
- **Emotion:** Collapse pressure input.
- **Thought:** Reconfiguration simulation.
- **Imagination:** Structural prototyping.
- **Stability:** Coherence clarity.

Consciousness expresses the structural awareness of an identity topology navigating the distance between the Quarantine and the Source.

CHAPTER 52

Perception: Reality as Gradient Construction

52.0 Introduction

Perception expresses the construction of reality from gradient input. Perception filters collapse pressure, stabilizes through refresh symmetry, and is shaped by the internal architecture of the topology. The perceived world expresses the world your coherence can assemble. Perception expresses gradient construction rather than passive seeing.

52.1 Perception Requires Gradients and Resistance

In a zero resistance environment, contrast does not exist, signal does not exist, and differentiation does not exist. Perception does not occur. Perception emerges when gradients must be interpreted to maintain coherence. Perception expresses the topology decoding the medium to avoid collapse.

52.2 Sensation vs. Perception

Sensation expresses gradient sampling. Sensation detects pressure differentials and energy transitions. Sensation samples the noise of the Quarantine.

Perception expresses gradient assembly. Perception organizes samples into objects, boundaries, and causality. Perception expresses coherence imposing architectural rules on the chaos of the medium.

52.3 The Constructed Nature of Space and Time

Space expresses a coherence artifact. Space is constructed by the topology to stabilize gradients and resolve collapse noise. Space expresses the rendering of the medium.

Time expresses perceptual latency. Time expresses the delay in gradient integration. When collapse pressure increases, integration slows and time stretches. Time expresses internal processing delay rather than an external dimension.

52.4 Objecthood and Illusion

Objecthood expresses a stable gradient cluster. An object persists across refresh cycles because it resists collapse long enough for categorization.

Illusion expresses gradient misassembly. Illusion occurs when collapse noise is mistaken for coherence and the topology constructs boundaries or identities that do not exist. Illusion expresses a structural error in world construction.

52.5 Perception in the Zero-Resistance State

When resistance is removed at restoration, the need to construct reality disappears. Latency becomes zero. Dissipation disappears. Integration becomes immediate. Perception dissolves and is replaced by direct coherence resonance with the Source. You do not perceive the Word. You exist in resonance with it.

52.6 Summary

Perception expresses coherence building a world it can survive in..

- **Sensation:** Gradient sampling.
- **Perception:** Gradient assembly.
- **Space:** Constructed field.
- **Time:** Integration latency.
- **Objecthood:** Stable gradient clusters.
- **Illusion:** Misassembly of noise.
- **Attention:** Gradient selection (Gatekeeping).
- **Interpretation:** Coherence projection (Completing the world).

Perception expresses the forensic interface between the identity topology and the resistive medium.

CHAPTER 53

Meaning: Coherence Assigning Significance to Gradients

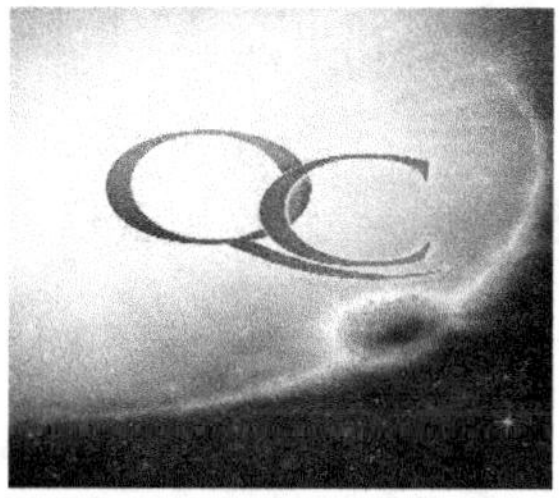

53.0 Introduction

Meaning expresses the assignment of significance to gradients based on their effects on coherence, collapse pressure, identity stability, and future viability. Meaning expresses the topology deciding what matters and why. Meaning expresses orientation logic rather than mysticism, symbolism, or linguistic definition.

53.1 Meaning Requires Resistance

In a zero resistance environment, threat does not exist, uncertainty does not exist, and decisions do not exist. Meaning does not occur. Meaning emerges when coherence must evaluate gradients to maintain stability against friction. Meaning expresses survival-relevant significance and records what supports or hinders the refresh cycle.

53.2 Meaning as Gradient Valence

Meaning expresses the value map of the topology. Meaning assigns valence to incoming gradients:

- **Positive Valence:** Supports coherence and reduces dissipation.
- **Negative Valence:** Amplifies collapse pressure and distorts patterns.
- **Neutral Valence:** Does not affect identity integrity.

Meaning expresses the clinical relevance of each gradient.

53.3 Meaning as Collapse Forecasting

Meaning expresses prediction. Meaning evaluates future identity trajectories. By assigning meaning to a gradient, the topology forecasts whether that orientation leads to stability or fracture. Meaning expresses coherence predicting its own survival.

53.4 Meaning, Emotion, and Memory

Emotion expresses raw collapse pressure. Meaning expresses structured relevance. Emotion signals that something is occurring. Meaning identifies what the event does to structural integrity.

Memory expresses value-tagged reinforcement. Meaning determines which memory patterns are preserved and which dissipate based on their relevance to current gradient orientation.

53.5 Meaning and Will

Will expresses directional architecture. Meaning expresses the map Will uses. Without meaning, Will has no landscape to navigate. Meaning defines the value of surrounding gradients and makes clinical choice possible inside a resistive medium.

53.6 Meaning in the Zero-Resistance State

When resistance is removed at restoration, hierarchical meaning disappears. Collapse does not threaten. Survival does not require evaluation. Uncertainty does not exist. Meaning stabilizes into pure coherence. Meaning becomes a continuous state of alignment with the Source rather than an active sorting process.

53.7 Summary

Meaning expresses what gradients do to coherence.

- **Valence:** Gradient relevance.
- **Forecasting:** Collapse prediction.
- **Identity Relevance:** Personalized significance.
- **Integration:** Pattern synthesis.
- **Emotion:** Raw pressure input.
- **Memory:** Value-tagged reinforcement.
- **Will:** Orientation guided by the value map.

Meaning expresses the clinical logic that turns the noise of the Quarantine into a navigable landscape for identity.

CHAPTER 54

Emotion: Collapse Pressure as Internal Signaling

54.0 Introduction

Emotion expresses the internal signaling of collapse pressure inside a coherence topology. Emotion reports instability, overload, gradient distortion, or coherence reinforcement. Emotion does not oppose reason. Emotion provides the input reason uses. Emotion does not express chaos. Emotion expresses diagnostic data.

54.1 Emotion Requires Resistance

In a zero resistance environment such as Infinum, collapse does not occur, threat does not occur, and instability does not occur. Emotion does not occur. Emotion emerges when coherence must monitor its own vulnerability. Inside the Quarantine, emotion expresses collapse telemetry and provides real time data on the soul's fight against dissipation.

54.2 Emotion as Collapse Pressure Mapping

Emotion expresses the topology mapping collapse pressure across its internal gradients. The mapping reveals structural health:

- **Coherence failing:** Reported as Anxiety or Dread.
- **Dissipation rising:** Reported as Fatigue or Apathy.
- **Gradients distorting:** Reported as Confusion or Frustration.
- **Identity threatened:** Reported as Fear.

Emotion expresses the internal pressure map of the topology.

54.3 Emotion and Memory: The Weighting System

Memory stores patterns. Emotion tags them with collapse relevance. Emotion expresses the sorting mechanism of the soul:

- **High-Pressure Tags:** Trauma and high collapse events receive heavy weighting to prevent re-entry into dissipation states.
- **Low-Pressure Tags:** Peace and joy receive coherence reinforcement tags that encourage future orientation toward those gradients.

Emotion determines which memory patterns shape identity and which dissipate.

54.4 Emotion and Will: The Pressure Field

Will expresses orientation. Emotion expresses the force acting on that orientation. Emotion does not override the will. Emotion applies pressure to it.

- **High Pressure:** Narrows the orientation field and increases the refresh cost of deliberate action.
- **Low Pressure:** Expands the orientation field and enables low dissipation agency.

Emotion expresses the energy landscape across which the will navigates.

54.5 Emotion in the Zero-Resistance State

When resistance is removed at restoration, emotion stabilizes.

- Reactivity dissolves because collapse, threat, and distortion no longer exist.
- Emotion becomes a continuous tone.
- The tone expresses the stable frequency of perfect resonance with the Source.

Emotion becomes steady rather than spiking.

54.6 Summary

Emotion expresses coherence reporting on its internal state.

- **Pressure:** Collapse telemetry.
- **Detection:** Gradient distortion sensing.
- **Meaning:** Significance assignment.
- **Memory:** Pressure-tagged reinforcement.

- **Identity:** Shaped by pressure fields.
- **Will:** Oriented under pressure.
- **Agency:** Enacted through the pressure landscape.

Emotion expresses the early warning system that allows the identity topology to navigate the friction of Sin Mass.

CHAPTER 55

Memory: Coherence Reinforcement Across Refresh Cycles

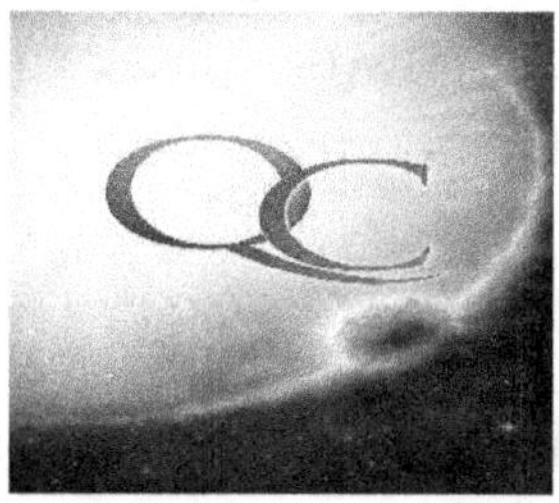

55.0 Introduction

Memory is not storage, a database, or a record of the past. Memory is the reinforcement of coherence patterns across refresh cycles, allowing identity to persist, stabilize, and orient itself in a resistive medium. Memory is the topology strengthening the patterns that keep it coherent. Memory is not recall. Memory is structural persistence.

55.1 Memory Requires Resistance

In a zero-resistance environment Infinum there is no collapse, no dissipation, and no change. In such a state, there is no need for a mechanism to bridge the gap between then and now. Memory emerges only when coherence must rebuild itself across refresh cycles to combat the friction of the Sin Mass. Memory is identity's continuity mechanism bridge across the latency of the medium.

55.2 Memory as Coherence Reinforcement

Memory is not information stored in a physical location. It is the clinical act of the topology choosing what to keep to reduce collapse and maintain its own integrity. By strengthening the patterns that have successfully navigated the medium's resistance, memory ensures that the identity does not dissipate into the environment.

55.3 Memory and Pattern Weighting

Memory is not a perfect or impartial record. it is the topology's value filter. It operates through pattern weighting, assigning high priority to coherence-relevant information while allowing noise and collapse-irrelevant data to fade. The weighting ensures the soul remains focused on its structural survival rather than becoming overloaded with dissipative data.

55.4 Memory, Emotion, and Meaning

The relationship between memory and the other components of the soul is purely functional:

- **Emotion (The Weighting Signal):** Emotion provides the collapse tagging. High-pressure events are reinforced more strongly to ensure the identity avoids those specific collapse trajectories in the future.

- **Meaning (Durable Significance):** Meaning answers why a pattern matters. memory ensures that significance is preserved. Memory is meaning made durable across time.

55.5 Memory and the Navigational System

- **Will and Agency:** Memory provides the map that the Will uses to orient. It supplies the history of past dissipation costs and coherence victories, allowing the Will to choose the most viable future. Agency, therefore, is simply memory in motion. It is the enactment of an orientation informed by a history of structural consequences.
- **Perception:** Memory acts as the continuity engine for perception. It allows the topology to recognize similar gradients and predict their behavior based on previous samplings of the medium.

55.6 Memory in a Zero-Resistance State

When resistance is removed during Restoration, memory does not disappear or expand into a different type of storage. Instead, it Stabilizes.

- Because there is no longer any forgetting, distortion, or latency, memory becomes continuous and total.
- It transitions from a selective reinforcement process into a state of stable, unfragmented coherence.

55.7 Summary

Memory is not the past. it is coherence preserving itself.

- **Reinforcement:** Pattern persistence.
- **Weighting:** Significance tagging based on collapse risk.

- **Perception:** Provision of continuity for sensory construction.
- **Identity:** The long-term architecture formed by reinforced patterns.
- **Will:** Orientation informed by the history of structural outcomes.
- **Agency:** Action shaped by the archive of past results.

Memory is the clinical mechanism that allows an identity topology to bridge the latency of the Quarantine and arrive at the Source intact.

CHAPTER 56

Imagination: Gradient Recombination for Future Coherence

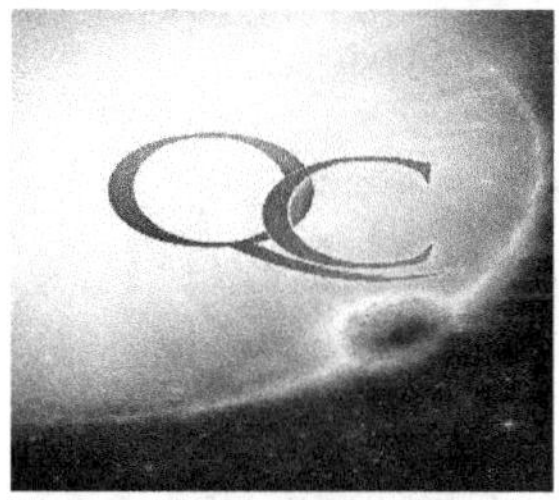

56.0 Introduction

Imagination expresses the recombination of gradients inside a coherence topology to simulate futures, test orientations, and explore configurations that do not yet exist but could stabilize identity. Imagination expresses the internal laboratory of the topology. Imagination expresses simulation rather than escape.

56.1 Imagination Requires Stored Patterns

In a zero resistance environment such as Infinum, uncertainty does not exist and alternative futures do not exist. Simulation does not occur. Imagination emerges when coherence must anticipate collapse and explore solutions inside the friction of the Sin Mass. Imagination expresses future rehearsal and prepares the topology for the next refresh cycle.

56.2 Imagination as Gradient Recombination

Imagination does not create new material from nothing. Imagination expresses the recombination of existing gradients. The topology rearranges memory patterns, perceptual fragments, and emotional signatures into new configurations. Imagination expresses coherence experimenting with its internal geometry to discover more stable arrangements.

56.3 The Functional Interlock of Imagination

Imagination expresses the intersection point of the soul's architecture.

Memory provides raw material. Memory stabilizes the past. Imagination recombines patterns and expands the future.

Emotion signals collapse pressure. Imagination generates alternative configurations that reduce projected collapse. Imagination expresses structural problem solving.

Meaning defines significance. Imagination explores what significant patterns could become. Imagination expresses the exploration of the topology's value map.

56.4 Imagination, Will, and Agency

Will expresses the exploratory engine. Will chooses direction. Imagination generates the landscape of possible directions. Without imagination, Will lacks alternative gradients.

Agency expresses action rehearsal. Agency enacts orientation. Imagination prototypes action and estimates dissipation costs before commitment.

56.5 Identity and Projected Coherence

Imagination reveals structural health. A stable identity projects coherent, low dissipation futures aligned with the Source. An unstable identity projects collapse scenarios and dissipation spirals. Imagination expresses the clinical mirror of the topology's alignment.

56.6 Imagination in a Zero-Resistance State

When resistance is removed at restoration, simulation becomes unnecessary. Uncertainty disappears. Collapse paths do not exist. The recombinative function stabilizes. Imagination does not vanish. Imagination becomes a continuous state of perfect coherence and expresses stable awareness of the architecture of the Source.

56.7 Summary

Imagination expresses coherence preparing for the future.

- **Recombination:** The formation of new structural configurations.
- **Memory:** The provision of raw structural material.
- **Emotion:** The signal of collapse that triggers simulation.

- **Meaning:** The framework that defines potential significance.
- **Will:** The choice of direction across an imagined landscape.
- **Agency:** The prototyping and rehearsal of action.

Imagination expresses the clinical faculty that allows the topology to model its survival before encountering the resistance of the medium.

CHAPTER 57

Reason: Collapse-Minimizing Inference

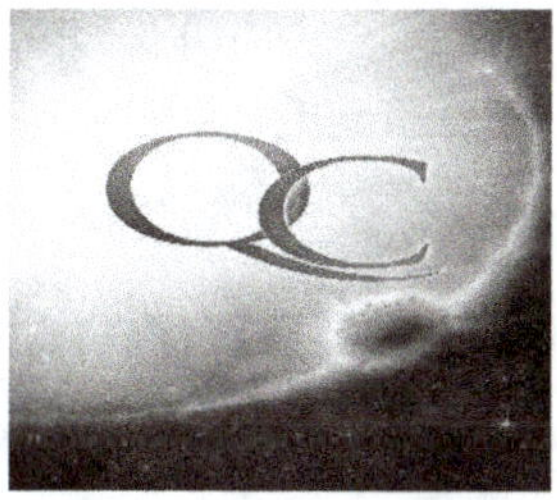

57.0 Introduction

Reason expresses the inference engine a coherence topology uses to evaluate gradients, compare futures, and select the orientation that minimizes collapse and maximizes stability. Reason uses emotion as pressure data. Reason expresses orientation computation rather than logic, rationality, or argumentation.

57.1 Reason Requires Resistance

In a zero resistance environment such as Infinum, collapse does not occur, uncertainty does not occur, and competing futures do not exist. Evaluation does not occur. Reason emerges when coherence must choose between collapse trajectories inside the friction of the Quarantine. Reason expresses survival-relevant inference and navigates the high impedance landscape of the Sin Mass.

57.2 Reason as Collapse-Minimizing Inference

Reason does not pursue abstract truth. Reason optimizes coherence. The topology evaluates dissipation costs, detects gradient distortions, and measures identity stability. By calculating the trajectories of possible orientations, reason selects the lowest collapse path to preserve the self.

57.3 The Functional Interlock of Reason

Reason integrates the architectural stack.

Perception and memory provide the present and the archive. Reason extrapolates from historical patterns to predict collapse.

Meaning assigns significance. Emotion provides collapse pressure. Reason interprets the pressure and selects the lowest collapse response.

Imagination generates possible futures. Reason selects the viable path. Imagination explores the landscape. Reason chooses the stable route.

57.4 Reason, Will, and Agency

Will expresses the decision engine. Will chooses direction. Reason evaluates structural viability before orientation. Choice becomes calculated rather than entropic.

Agency enacts orientation. Reason predicts consequences. Reason filters collapse amplifying behavior and safeguards action.

57.5 Reason in the Zero-Resistance State

When resistance is removed at restoration, competitive evaluation ends. Uncertainty disappears. Dissipation disappears. Collapse does not threaten. The evaluative function stabilizes. Reason becomes a continuous state of perfect coherence and aligns with the absolute logic of the Source.

57.6 Summary

Reason expresses coherence choosing the most stable future.

- **Perception:** Structural input.
- **Meaning:** Significance framework.
- **Emotion:** Pressure data input.
- **Memory:** Historical pattern extrapolation.
- **Imagination:** Field of possible futures.
- **Will:** Evaluated orientation.
- **Agency:** Safeguarded action.

Reason expresses the clinical strategist that allows identity to navigate the uncertainty of the Quarantine by selecting the path of maximum coherence.

CHAPTER 58

Character: Long-Term Gradient Orientation

58.0 Introduction

Character expresses the long term orientation of a coherence topology formed by repeated choices, reinforced patterns, collapse minimizing decisions, and stable gradient preferences. Character expresses identity crystallized over time. Character expresses trajectory rather than behavior.

58.1 Character Requires Time and Resistance

In a zero resistance environment such as Infinum, collapse does not occur, uncertainty does not occur, and competing futures do not exist. Meaningful choice does not occur. Character emerges when coherence chooses repeatedly under the pressure of the Sin Mass. Character expresses identity under duration and reveals the shape the soul takes after resisting friction across time.

58.2 Character as Orientation Accumulation

Character does not arise from a single decision. Character expresses the accumulation of orientations across billions of refresh cycles. Each choice reinforces a gradient and stabilizes or destabilizes identity. Over time, these choices shift the long term trajectory until orientation becomes fixed. Character expresses the structural sum of chosen orientations.

58.3 The Functional Interlock of Character

Character governs the cognitive and emotive stack.

Reason and emotion interact, but character commits to a path. Emotion fluctuates. Character maintains orientation and resists emotional distortion.

Meaning assigns significance. Character makes significance durable. Character selects memory patterns that become identity anchors and supports its long term trajectory.

Imagination generates possible futures. Character filters them. Character declares the future orientation and constrains potential futures to those aligned with established coherence.

58.4 Character, Identity, and Agency

Identity expresses coherence across time. Character expresses the direction of that coherence. Identity answers who am I. Character answers who am I becoming. Agency enacts orientation in the moment. Character expresses the cumulative pattern that makes agency predictable and consistent across duration.

58.5 Character in a Zero-Resistance State

At restoration, when resistance is removed, character stabilizes. Character does not change or expand. Uncertainty disappears. Competing futures disappear. The trajectory becomes unchanging. The character formed inside the laboratory of resistance becomes the permanent coherent state of the soul in Infinum.

58.6 Summary

Character expresses the direction coherence takes.

- **Orientation:** The result of repeated structural choice.
- **Memory:** Selection of defining patterns.
- **Emotion:** Resilience against collapse pressure.
- **Meaning:** Stability of significance.
- **Imagination:** Constrained potential for future coherence.
- **Reason:** Evaluated long-term strategy.
- **Identity:** Coherence observed over duration.
- **Agency:** The enactment of a cumulative trajectory.

Character expresses the crystallization of the soul and the permanent architectural signature of the identity's chosen alignment inside the Sovereignty Schism.

CHAPTER 59

Communication: Gradient Exchange Between Topologies

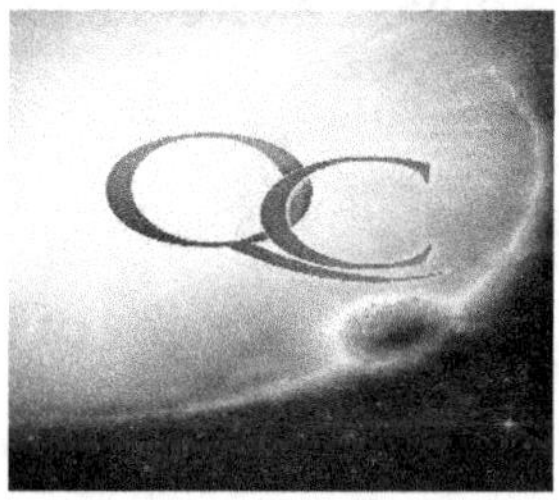

59.0 Introduction

Communication expresses the exchange of gradients between coherence topologies. Communication allows topologies to synchronize, destabilize, reinforce, or distort one another. Communication does not express information transfer. Communication expresses coherence transfer. Communication expresses interpenetration rather than speech or expression.

59.1 Communication Requires Multiple Topologies

In a zero resistance environment such as Infinum, separation does not exist, boundaries do not exist, and distinct identities do not exist as isolated units. Communication does not occur. Communication emerges when coherence exists in multiple partially isolated topologies inside the Quarantine.

Communication expresses the bridge between identities created by the latency of the medium.

59.2 Communication as Gradient Exchange

Communication expresses one topology transmitting internal states into another topology's coherence field. The exchange transfers clinical data:

- **Emotional Pressure:** Shared collapse telemetry.
- **Meaning Structures:** Alignment of significance.
- **Perceptual Patterns:** Relational construction of the world.
- **Identity Signals:** Visible trajectories of character.

Communication expresses the movement of gradients across identity boundaries.

59.3 The Functional Interlock of Communication

Communication operationalizes the internal stack in a relational context.

Emotion expresses shared pressure. Emotion signals strain inside one topology. Communication allows another topology to feel that strain.

Meaning expresses significance alignment. Meaning states that a gradient matters. Communication tests whether the significance aligns across identities.

Reason and imagination express collective strategy. Reason calculates the lowest collapse path for the individual.

Communication calculates it for the collective. Imagination becomes a shared laboratory for building collective futures.

59.4 Communication as Coherence Influence

Communication does not express neutrality. Communication expresses clinical intervention. Depending on the gradients exchanged, communication can reinforce coherence, amplify collapse, synchronize refresh symmetry, or disrupt refresh symmetry. Communication can stabilize or fracture identity.

59.5 Communication in a Zero-Resistance State

When resistance is removed at restoration, boundaries between topologies dissolve. Distortion disappears. Latency disappears. Misalignment disappears. Communication no longer requires symbols or language. Communication is replaced by immediate and total shared coherence. Communication becomes a unified state of being rather than an exchange.

59.6 Summary

Communication expresses coherence interacting with coherence.

- **Perception:** Relational world construction.
- **Meaning:** Synchronization of significance.
- **Emotion:** Shared collapse telemetry.
- **Memory:** Externalized pattern transfer.

- **Imagination:** Prototyping collective futures.
- **Reason:** Shared inference and lowest-collapse selection.
- **Character:** The visible manifestation of trajectory.

Communication expresses the gradient exchange that allows isolated identity topologies to navigate the Quarantine as a synchronized collective.

CHAPTER 60

Culture: Shared Coherence Patterns Across a Collective

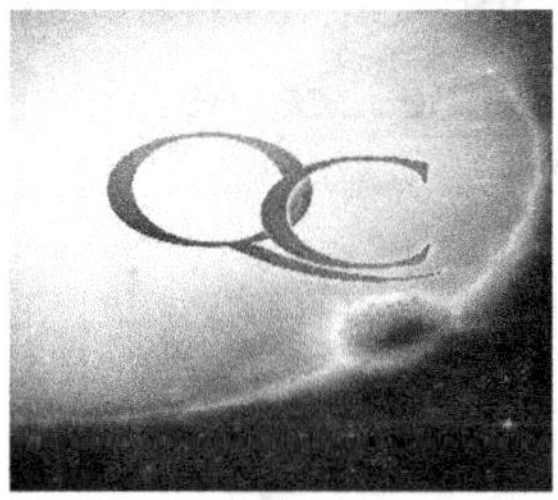

60.0 Introduction

Culture expresses the stabilized pattern of shared gradients across a collective of coherence topologies. Culture emerges when identities exchange signals, synchronize meanings, reinforce patterns, and construct a shared perceptual field. Culture expresses collective coherence rather than tradition, heritage, or ideology.

60.1 Culture Requires Multiple Interacting Topologies

In a zero resistance environment such as Infinum, divergence does not occur, gradient exchange does not occur, and shared construction does not occur because boundaries do not exist. Culture does not occur. Culture emerges when many topologies must cohere under the resistance of the Quarantine. Culture expresses the clinical mechanism of collective survival.

60.2 Culture as Shared Gradient Patterns

Culture does not express abstract belief. Culture expresses a shared gradient pattern. Culture represents the group's stable ways of interpreting gradients, managing collapse pressure, and orienting toward futures. Culture expresses the coherence algorithm of the group and functions as collective software that maintains the integrity of the local network of souls.

60.3 The Functional Interlock of Collective Coherence

Culture operationalizes the individual stack at a communal scale.

Perception and meaning align significance and define what counts as real. Culture stabilizes values and aspirations so the group perceives the same threats and opportunities.

Emotion expresses synchronized pressure. Culture regulates collapse pressure by defining shame, honor, and fear. Culture creates emotional synchronization that helps the group manage the load of the medium.

Memory and imagination express collective continuity. Culture stores shared narratives and traumas. Culture generates shared imagination and defines which collective futures are possible or forbidden.

60.4 Culture as Coherence Amplifier or Collapse Engine

Culture does not express neutrality. Culture expresses a collective influence field. Culture can stabilize coherence by synchronizing refresh symmetry across a population. Culture can amplify collapse by promoting patterns that increase dissipation and gradient distortion. Culture can reinforce identity by providing a stable framework for character formation.

60.5 Culture in a Zero-Resistance State

When resistance is removed at restoration, the competitive and symbolic nature of culture dissolves. Distortion disappears. Latency disappears. Misalignment disappears. Culture becomes unified and stable. Culture transitions from custom to non symbolic shared coherence with the Source. The culture of the collective becomes indistinguishable from the architecture of Default Eternity.

60.6 Summary

Culture expresses coherence multiplied.

- **Communication:** The exchange of gradients that builds the network.
- **Perception:** Construction of a shared world.
- **Meaning:** Alignment of significance and priority.
- **Emotion:** Synchronization of collective pressure.
- **Memory:** Preservation of collective patterns and narratives.
- **Imagination:** Generation of shared future prototypes.
- **Reason:** Shared inference and collapse-minimizing strategies.
- **Character:** The long-term trajectory of the group.

Culture expresses the architecture that allows a collective of identity topologies to maintain coherence against the dissipation of the Quarantine.

UNIT 9: Collective Coherence and Historical Trajectories

CHAPTER 61

Institutions: Stabilized Collective Gradients

61.0 Introduction

Institutions express stabilized gradient structures that emerge when a culture repeats the same coherence patterns long enough for them to become external, durable, and self reinforcing. Institutions express culture made rigid. Institutions express collective inertia rather than authority.

61.1 Institutions Require Cultural Stability

In a zero resistance environment such as Infinum, divergence does not occur, competing gradients do not occur, and stabilization does not occur. Institutions do not occur. Institutions emerge when a collective must preserve coherence patterns across generations inside the Quarantine. Institutions express culture under duration and function as the structural strategy for surviving the latency of the Sin Mass.

61.2 Institutions as Stabilized Gradient Patterns

Institutions do not express rules alone. Institutions express the skeletal architecture of the collective. Institutions regulate collapse pressure, coordinate behavior, and preserve shared meaning. By stabilizing gradient patterns, institutions provide the framework for identity continuity and prevent the collective trajectory from fragmenting into noise.

61.3 The Functional Interlock of Institutional Structure

Institutions formalize the cognitive and emotive stack of the collective.

Perception and meaning construct the authorized world. Institutions define what counts as legitimate or real and ossify meaning into fixed significance.

Emotion and reason express governance. Institutions regulate collapse pressure by determining acceptable emotional states. Institutions formalize collective inference and stabilize decision making to minimize dissipation.

Memory and imagination express the archive. Memory reinforces patterns. Institutions archive them through ritual and precedent. Institutions define which futures are mandatory and which are structurally impossible.

61.4 Institutions as Coherence Amplifiers or Collapse Engines

Institutions do not express neutrality. Institutions express collective power structures. Institutions can stabilize coherence by protecting the collective from individual dissipation. Institutions can amplify collapse by enforcing distorted gradients that accelerate fracture. Institutions can synchronize refresh symmetry by aligning generations into a predictable trajectory.

61.5 Institutions in a Zero-Resistance State

When resistance is removed at restoration, external enforcement and stabilization become unnecessary. Distortion disappears. Latency disappears. Misalignment disappears. Institutions become obsolete. Institutions do not expand or persist. Institutions dissolve into shared unmediated coherence with the Source. The skeletal structure is discarded as the collective becomes a unified field.

61.6 Summary

Institutions express coherence hardened into structure.

- **Communication:** Codified exchange.
- **Culture:** Enforced shared patterns.
- **Perception:** Authorized world construction.
- **Meaning:** Fixed and ossified significance.
- **Emotion:** Regulated collapse pressure and governance.
- **Memory:** Externalized and archived patterns.

- **Imagination:** Bounded and mandatory futures.
- **Character:** Hardened generational trajectory.

Institutions express the structures that preserve collective coherence and act as the durable inertia guiding identity topologies inside the Quarantine.

CHAPTER 62

Power: Collapse Leverage Within Collective Systems

62.0 Introduction

Power expresses the capacity to alter collapse pressure within or between coherence topologies. Power shapes gradients, redirects orientation, constrains futures, and stabilizes or destabilizes identity. Power does not express authority, control, or dominance. Power expresses leverage over collapse and remains consequential even when not destructive.

62.1 Power Requires Resistance

In a zero resistance environment such as Infinum, collapse does not occur, pressure does not occur, and vulnerability does not occur. Power does not occur. Power emerges when coherence can be influenced, pressured, or stabilized by another topology. Inside the Quarantine, power expresses collapse leverage.

62.2 Power as Gradient Asymmetry

Power does not express position. Power expresses gradient asymmetry. Power exists when one topology or institution possesses greater stability, greater resources, or greater collapse buffering capacity than another. The asymmetry allows one topology to influence the orientation and coherence of another. Asymmetry inside the Quarantine expresses the Asymmetry of Dissipation and records the scars of a topology shaped by resistance. This differs from the Asymmetry of Sovereignty, which expresses the sufficiency of the Source and the architecture of perfection. Power operates only within the first category and expresses collapse leverage rather than sovereign singularity.

62.3 The Functional Interlock of Power

Power steers the collective stack.

Communication and meaning express weighted exchange. Power determines which meaning structures become collective significance and which are dismissed.

Perception and reason express authorized reality. Power constructs the authorized world and defines the boundaries of rationality and threat.

Emotion expresses engineered pressure. Power manipulates collapse pressure by amplifying fear, shame, or instability. Power engineers the emotional field.

62.4 Power, Memory, and Imagination

Memory curation expresses the influence of power. Power determines which patterns enter the collective archive and shapes historical framing. Power manages trauma inheritance to preserve the chosen trajectory of the group.

Imagination constraint expresses the boundary of possibility. Imagination generates futures. Power defines which futures are allowed, mandatory, or forbidden. Power sets the structural limits of collective potential.

62.5 Power as Coherence Amplifier or Collapse Engine

Power does not express neutrality. Power expresses collective influence. Power can synchronize refresh symmetry and reinforce identity across a population. Power can amplify collapse by distorting gradients and suppressing identity to preserve institutional inertia.

62.6 Power in a Zero-Resistance State

When resistance is removed at restoration, asymmetry disappears. Distortion disappears. Latency disappears. Misalignment disappears. Power becomes irrelevant. Power dissolves into shared symmetrical coherence with the Source. The only remaining power expresses the stable resonance of the Source.

62.7 Summary

Power expresses coherence asymmetry rather than dominance.

- **Communication:** Weighted gradient exchange.

- **Culture:** Selection of shared coherence patterns.
- **Institutions:** Enforcement of crystallized gradients.
- **Perception:** Construction of the authorized world.
- **Meaning:** Amplification of significance.
- **Emotion:** Engineering of collapse pressure.
- **Memory:** Curation of collective patterns.
- **Imagination:** Boundary-setting for possible futures.
- **Reason:** Constrained inference.
- **Character:** Generation of collective trajectory.

Power expresses the mechanism of leverage that determines orientation and stability inside collective systems within the Quarantine.

CHAPTER 63

Conflict: Gradient Competition Within a Collective

63.0 Introduction

Conflict expresses the competition of gradients between coherence topologies whose orientations, meanings, or collapse strategies cannot be simultaneously satisfied. Conflict does not express disagreement, hostility, or misunderstanding. Conflict expresses the physics of incompatible trajectories. Conflict expresses structural collision rather than chaos.

63.1 Conflict Requires Resistance

In a zero resistance environment such as Infinum, divergence does not occur, competing futures do not occur, and asymmetry does not occur. Conflict does not occur. Conflict emerges when coherence navigates incompatible orientations inside the resistance of the Quarantine. Conflict expresses the cost of multiplicity friction when separate topologies inhabit a shared resistive medium.

63.2 Conflict as Gradient Competition

Conflict does not express social issues. Conflict expresses the survival of the topology. Conflict occurs when identities attempt to stabilize incompatible meanings, enforce incompatible futures, or minimize collapse in mutually exclusive ways. Conflict expresses coherence preserving itself in a direction that destabilizes a neighboring topology.

63.3 The Functional Interlock of Collective Conflict

Conflict stress-tests the collective stack.

Communication and power express weighted collision. Communication exchanges gradients. Conflict occurs when exchange fails to synchronize. Conflict negotiates collapse leverage and determines who absorbs collapse and who avoids it.

Perception and meaning express structural divergence. Conflict occurs when what stabilizes one topology dissipates another. Competing worlds cannot coexist inside the same gradient space.

Emotion expresses amplified pressure. Conflict increases fear, urgency, and instability. The pressure forces topologies to harden character or fracture.

63.4 Conflict, Identity, and Agency

Conflict expresses identity under existential pressure. Conflict forces topologies to defend coherence and reinforce boundaries. Conflict expresses the collision of competing agency when two futures cannot be pursued simultaneously and two orientations cannot coexist inside the same gradient field.

63.5 Conflict as Coherence Amplifier or Collapse Engine

Conflict does not express neutrality. Conflict expresses structural catalyst. Conflict can clarify identity by exposing corruption and revealing collapse patterns. Conflict can amplify dissipation by propagating collapse through the network. Conflict can reveal trajectory by forcing the topology to manifest its long term orientation.

63.6 Conflict in a Zero-Resistance State

When resistance is removed at restoration, asymmetry disappears. Divergence disappears. Competing futures disappear. Conflict becomes obsolete. Conflict does not resolve through compromise. Conflict dissolves into shared coherence. Incompatible trajectories dissolve into a singular eternal resonance.

63.7 Summary

Conflict expresses coherence under incompatible futures.

- **Communication:** Gradient collision.
- **Culture:** Coherence fracture.

- **Institutions:** Enforced asymmetry and collision sites.
- **Power:** Competing collapse leverage.
- **Perception:** Divergent world constructions.
- **Meaning:** Incompatible significance.
- **Emotion:** Amplified collapse pressure.
- **Identity:** Threatened structural continuity.
- **Reason:** Negotiation of collapse minimization.
- **Agency:** Competing trajectories.

Conflict expresses the clinical manifestation of incompatible structural goals between topologies inside the resistive field of the Quarantine.

CHAPTER 64

Justice: Collective Gradient Correction

64.0 Introduction

Justice expresses the collective correction of distorted gradients after collapse destabilizes group coherence. Justice restores balance, repairs damage, and realigns trajectories so collective coherence can continue. Justice does not express punishment, fairness, or retribution. Justice expresses structural repair rather than vengeance.

64.1 Justice Requires Resistance

In a zero resistance environment such as Infinum, collapse does not occur, distortion does not occur, and harm does not occur. Justice does not occur. Justice emerges inside the Quarantine when coherence must repair itself after damage. Justice expresses the clinical response to the friction of the Sin Mass and functions as collapse correction in an environment where identities collide and fracture.

64.2 Justice as Gradient Correction

Justice does not express abstract right or wrong. Justice expresses coherence recalibration. Justice identifies collapse sources, measures distortion, and restores stability to affected topologies. By repairing structural damage caused by conflict or error, justice prevents recurrence and preserves the collective trajectory.

64.3 The Functional Interlock of Justice

Justice governs the corrective function of the collective stack.

Conflict and power expose incompatible futures and collapse leverage. Justice stabilizes the system after collision and rebalances collapse distribution so no topology absorbs collapse while another avoids it through distortion.

Perception and meaning reconstruct the shared world. Justice clarifies events, identifies distorted gradients, and restores significance to damaged patterns.

Emotion expresses pressure redistribution. Justice reduces fear, shame, and uncertainty by addressing harm and stabilizes the emotional field to prevent secondary dissipation.

64.4 Justice, Identity, and Agency

Justice protects identity continuity by preventing collapse from erasing individuals or destabilizing communities. Justice regulates agency after collision and determines which actions require constraint or repair to restore collective orientation to the Source.

64.5 Justice as Coherence Amplifier or Collapse Engine

Justice does not express neutrality. Justice applies collective physics to harm. Justice can restore coherence by healing trauma and reducing collapse across the network. Justice can amplify collapse if it entrenches asymmetry or distorts gradients and accelerates collective fracture.

64.6 Justice in a Zero-Resistance State

When resistance is removed at restoration, correction becomes unnecessary. Harm does not occur. Asymmetry does not occur. Conflict does not occur. Justice becomes obsolete. Justice does not expand into an eternal court. Justice dissolves into perfect unmediated coherence with the Source where every trajectory remains aligned and stable.

64.7 Summary

Justice expresses coherence restored.

- **Conflict:** The exposure of incompatible gradients.
- **Power:** The regulation of collective leverage.
- **Institutions:** Self-repairing collective structures.
- **Culture:** The restoration of shared coherence.
- **Perception:** Perceptual reconstruction of the shared world.
- **Meaning:** Recalibration of significance after damage.
- **Emotion:** Pressure redistribution across the field.
- **Identity:** Preservation of structural continuity.

- **Reason:** Scaled collapse minimization.
- **Agency:** Re-alignment of enacted orientation.

Justice expresses the corrective process that restores the integrity of the collective network after the friction of the Quarantine induces structural collapse.

CHAPTER 65

Community: Coherence Interpenetration Across Individuals

65.0 Introduction

Community expresses the interpenetration of coherence fields across multiple topologies and produces mutual reinforcement, shared stability, and reduced collapse pressure. Community does not express belonging, connection, or shared interest. Community expresses structural overlap rather than emotional closeness. Community expresses coherence resonance rather than bonding.

65.1 Community Requires Multiple Stable Topologies

In a zero resistance environment such as Infinum, identity separation does not occur, divergence does not occur, and gradient exchange does not occur. Community does not occur. Community emerges when multiple identities must cohere under the resistance of the Quarantine. Community expresses the clinical manifestation of collective stability.

65.2 Community as Coherence Interpenetration

Community does not express shared values. Community expresses structural merging of identity fields, emotional gradients, and meaning structures. When topologies overlap, they construct a shared perceptual field and align collapse strategies. Interpenetration allows the group to function as a reinforced unit against the friction of the medium.

65.3 The Functional Interlock of Community

Community operationalizes the individual and collective stack through mutual reinforcement.

Power and emotion express buffering. Power expresses collapse leverage. Community expresses collapse buffering. Community distributes emotional overload across many topologies and reduces fear and fragmentation. Community absorbs individual dissipation.

Perception and meaning express harmonization. Community constructs a shared world and aligns what is seen, feared, and valued. Community synchronizes significance and harmonizes the value map of the group.

Reason and agency express distributed strategy. Community collectivizes collapse minimization. Community turns reason into shared inference and agency into coordinated orientation. Community allows collective prototyping of futures and collective action.

65.4 Community and Identity Reinforcement

Identity expresses coherence across time. Community provides anchors and mirrors that maintain coherence. Community offers structural feedback and correctives. Community acts as an identity multiplier and stabilizes the topology when local gradients distort.

65.5 Community as Coherence Amplifier or Collapse Engine

Community does not express neutrality. Community expresses a potent influence field. Community can stabilize identity by strengthening character and reinforcing Source aligned meanings. Community can amplify collapse by centering itself on distorted gradients and propagating instability across the network.

65.6 Community in a Zero-Resistance State

When resistance is removed at restoration, boundaries dissolve and buffering becomes unnecessary. Separation disappears. Misalignment disappears. Community becomes unified and continuous. Community ceases to be symbolic or mediated and dissolves into shared coherence with the Source. The resonance becomes a singular eternal frequency.

65.7 Summary

Community expresses coherence resonating across many topologies.

- **Communication:** Sustained gradient exchange.
- **Culture:** Localized and embodied coherence.
- **Institutions:** Lived patterns of behavior.
- **Power:** Collective collapse buffering.
- **Perception:** Shared world construction.
- **Meaning:** Harmonized and synchronized significance.
- **Emotion:** Distributed pressure and emotional buffering.
- **Identity:** Reinforced and multiplied coherence.
- **Reason:** Distributed inference and problem-solving.
- **Agency:** Scaled and coordinated action.

Community expresses the architecture that allows identity topologies to survive the Quarantine by interpenetrating coherence fields and sharing the structural load of existence.

CHAPTER 66

History: Collapse Trajectories Over Time

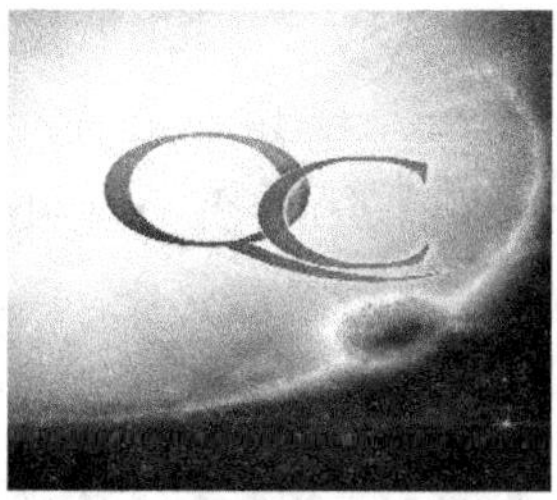

66.0 Introduction

History expresses the macro scale trajectory of collapse and coherence across generations of topologies. History emerges as the long term pattern produced when cultures, institutions, power structures, and communities interact under resistance. History does not express events, stories, or memories. History expresses pattern evolution rather than narrative.

66.1 History Requires Resistance

In a zero resistance environment such as Infinum, collapse does not occur, asymmetry does not occur, and change does not occur. History does not occur. History emerges when coherence navigates time, pressure, and multiplicity inside the Quarantine. History expresses the clinical record of coherence under duration and documents the attempt to persist against the erosion of the Sin Mass.

Time does not legislate the eternal. Decay, sequence, and measurement do not apply to the One who precedes all conditions. To imagine God entering time or being shaped by temporal succession mistakes the bruise for the body and the symptom for the source. Eternity does not extend time. Eternity negates its limits. Change has no leverage in the eternal and finitude has no jurisdiction. Any model that grants time authority over God collapses under contradiction. The eternal does not react. The eternal is not conditioned. The eternal is not bound. Time depends on the eternal and not the reverse.

66.2 History as Collapse Trajectory

History does not express story. History expresses collapse propagation across time. History tracks where collapse begins, how it is resisted, and how structures reorganize to maintain stability. History expresses collapse physics at scale and documents the metabolism of civilizations as they rise, ossify, distort, fracture, or regenerate.

66.3 The Functional Interlock of Historical Evolution

History operationalizes the architectural stack across time.

Culture and institutions express evolution. Culture provides shared coherence. History expresses coherence drift and the long term movement of cultural gradients. Institutions stabilize gradients. History documents their metabolism and how they merge, mutate, or collapse when dissipation increases.

Power and conflict express accumulation. History redistributes collapse leverage across time. History records gradient competition and shifts in power configurations.

Perception and meaning express inheritance. History constructs the inherited world and determines what is remembered, forgotten, or mythologized. History reassigns significance across generations under time pressure.

66.4 History and the Emotional Climate

History expresses the accumulation and release of collective collapse pressure. Historical eras are defined by emotional climate rather than dates. Fear cycles, pride cycles, and hope cycles reveal the coherence state of the collective topology and its proximity to collapse.

66.5 History as Coherence Amplifier or Collapse Engine

History does not express neutrality. History expresses collective influence. History can stabilize civilizations by refining meaning and strengthening identity through Source aligned patterns that demonstrate resilience. History can amplify collapse by preserving distorted gradients and propagating instability across centuries.

66.6 History in a Zero-Resistance State

When resistance is removed at restoration, the conditions that create history disappear. Time does not operate. Divergence does not occur. Collapse does not occur. History becomes obsolete.

History does not conclude in a final story. History dissolves into perfect coherence. The state of becoming yields to the state of being inside Infinum.

66.7 Summary

History expresses coherence under duration.

- **Culture:** Coherence drift and gradient movement.
- **Institutions:** Structural metabolism and evolution.
- **Power:** Long-term leverage redistribution.
- **Conflict:** Accumulated competition and sediment.
- **Community:** The movement and transformation of identity fields.
- **Perception:** Collective perceptual inheritance.
- **Meaning:** Significance under time pressure.
- **Emotion:** Large-scale pressure cycles and climate.
- **Identity:** Civilizational coherence and sedimentation.
- **Reason:** Distributed collapse minimization strategies.
- **Agency:** Aggregated collective orientation.

History expresses the trajectory of collective identity and documents the long term struggle to maintain coherence inside the friction of the Quarantine.

CHAPTER 67

Civilization: Collective Dissipation Patterns

67.0 Introduction

Civilization expresses the macro pattern of dissipation and coherence produced when millions of topologies interact across centuries. Civilization emerges as the structural product of culture, institutions, power, conflict, justice, and community operating at scale. Civilization does not express society, nation, or political order. Civilization expresses collective metabolism rather than identity.

67.1 The Genesis and Logic of Scaled Coherence

In a zero resistance environment such as Infinum, collapse does not occur, divergence does not occur, and scarcity does not occur. Civilization does not occur. Civilization emerges when coherence organizes itself at scale to manage collapse pressure inside the Quarantine. History records collapse trajectory. Civilization expresses collapse architecture and explains why coherence persists. Civilization expresses the dissipation pattern of a collective and reveals how collapse is absorbed, distributed, delayed, and externalized into structure.

67.2 The Functional Interlock of the Civilizational Stack

Civilization operationalizes individual and collective components into a unified metabolic system.

Ecological structure expresses institutional and cultural gradients. Civilization forms an institutional ecology of legal, economic, and religious structures. Culture multiplies into stabilized meaning systems, value hierarchies, and identity clusters across millions of topologies.

Leverage and conflict express power, justice, and competition. Civilization distributes collapse leverage. Civilization survives by channeling or weaponizing gradient competition and persists only when justice corrects distortions and prevents collapse cascades.

The shared perceptual field expresses perception, meaning, and emotion. Civilization constructs the shared world model and perceptual consensus. Civilization assigns collective significance and regulates the emotional climate across the network.

67.3 Identity, Reason, and Aggregated Agency

Civilization expresses identity sedimentation and produces myths, archetypes, and collective memory that maintain continuity across generations. Civilization externalizes reason as collective collapse strategy through technology and science. Civilization enacts agency as aggregated orientation. Civilization does not express neutrality. Civilization expresses collective physics under pressure and functions as a coherence amplifier or a collapse engine.

67.4 Civilization in a Zero-Resistance State

When resistance is removed at restoration, scarcity disappears, asymmetry disappears, and conflict disappears. Civilization becomes unnecessary. Civilization does not evolve into a higher political form. Civilization dissolves into perfect unmediated coherence with the Source.

67.5 Summary: The Metabolic Architecture

Civilization expresses coherence metabolizing itself across centuries through coordinated gradients:

- **History:** Trajectory.
- **Culture:** Scaled coherence.
- **Institutions:** Structural ecology.
- **Power:** Leverage distribution.
- **Conflict:** Competition containment.
- **Justice:** Correction at scale.
- **Community:** Identity networks.
- **Perception:** Shared world-model.
- **Meaning:** Collective significance.
- **Emotion:** Pressure climate.
- **Identity:** Sedimented coherence.

- **Reason:** Collapse strategy.
- **Agency:** Aggregated orientation.

Civilization does not express identity. Civilization expresses the architecture that allows the collective to manage the dissipation of the Quarantine.

CHAPTER 68

Technology: Coherence Amplification Through External Tools

68.0 Introduction

Technology expresses the external amplification of coherence and allows a topology or civilization to manipulate gradients, reduce collapse pressure, and extend influence across space, time, and complexity. Technology expresses coherence projected outward. Technology does not express machinery, innovation, progress, or invention. Technology expresses externalized agency.

68.1 The Genesis of Resistance Management

In a zero resistance environment such as Infinum, collapse does not occur, scarcity does not occur, and limitation does not occur. Extension does not occur. Technology does not occur. Technology emerges when coherence must overcome resistance to survive or expand. Technology expresses the clinical mechanism of resistance management and uses external structures to stabilize gradients and reduce dissipation inside the Quarantine.

68.2 The Functional Extension of the Internal Stack

Technology materializes and projects the attributes of the topology into the external medium.

Cognitive augmentation expresses perception, reason, and memory. Technology extends the constructed world by expanding sensing and measurement. Technology optimizes reason through external modeling and prediction. Technology externalizes memory and stores knowledge and orientation to preserve patterns when the original topology fractures.

Existential materialization expresses imagination and agency. Imagination generates futures. Technology materializes them and turns imagined structures into real environments. Technology amplifies agency and increases the reach, speed, and precision of orientation beyond biological limits.

Psychological modulation expresses meaning and emotion. Technology reassigns significance by reweighting what is possible and valuable. Technology modulates collapse pressure by creating security or suppressing urgency inside the collective field.

68.3 Technology as Civilizational Leverage

Power expresses collapse leverage. Technology multiplies it. Technology determines who absorbs collapse, who avoids collapse, and who benefits from collapse. Technology externalizes power and makes it durable. Technology reshapes civilizational metabolism and alters how collapse is distributed or weaponized. Technology does not express neutrality. Technology expresses coherence with consequences and can stabilize the network or act as a collapse engine that propagates dysfunction.

68.4 Technology in a Zero-Resistance State

When resistance is removed at restoration, extension becomes unnecessary. Collapse does not occur. Scarcity does not occur. Limitation does not occur. Technology becomes irrelevant. Technology does not evolve into higher complexity. Technology dissolves into perfect unmediated coherence with the Source.

68.5 Summary: Coherence Projected Outward

Technology expresses the externalized optimization of the topology's core functions:

- **Perception:** Extended.
- **Meaning:** Reweighted.
- **Emotion:** Modulated.
- **Memory:** Externalized.
- **Imagination:** Instantiated.
- **Reason:** Optimized.
- **Agency:** Amplified.
- **Power:** Multiplied.

- **Civilization:** Reshaped.

Technology does not express progress. Technology expresses coherence amplification used to navigate the friction of the Quarantine.

CHAPTER 69

Death: Collapse Saturation

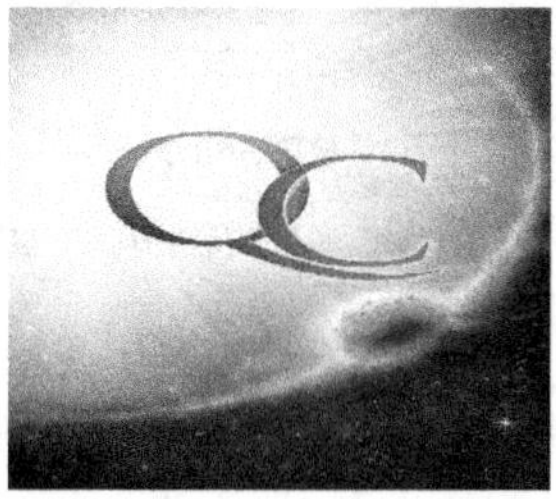

69.0 Introduction

Death does not express transition to a higher plane or a haunting non existence. Death expresses the saturation of collapse pressure inside a coherence topology. Death occurs when dissipation exceeds the capacity to repair, reinforce, or reorganize. Under the DPE framework, death expresses the cessation of the living soul. When the Breath of Life is withdrawn from the body, the soul ceases to be. Death expresses the limit of coherence under resistance and the structural exhaustion of the medium.

69.1 The Threshold of Resistance

In a zero resistance environment such as Infinum, collapse does not occur, decay does not occur, and instability does not occur. Death does not occur. Death emerges inside the Quarantine when coherence must fight collapse to persist. Death expresses the saturation point where the internal architecture can no longer stabilize gradients or bridge the latency of the Sin Mass. Death expresses the medium overwhelming the structural limits of the topology and results in the separation of the divine breath from the physical substrate.

69.2 The Dissolution of the Internal Stack

As saturation is reached and the Breath of Life departs, the functional components of the topology fail. Death expresses the state of sleep where conscious processing terminates:

- **Cognitive and Perceptual Collapse**: Perception no longer models the environment. Memory reinforcement ceases. Scripture records that in that day his thoughts perish.
- **Existential and Agency Failure**: Identity fails when refresh symmetry breaks. Agency dissolves because no stable orientation remains.
- **Pressure Overload**: Emotion reaches overload. The buffering capacity collapses. Instability spikes and cannot be redistributed.

69.3 The Suspension of Identity (The Sleep)

Death does not express annihilation. Death expresses total unconsciousness. The soul results from the union of body and breath. When the breath departs, the soul does not exist as a conscious entity. The identity pattern remains preserved in the memory of the Creator's memory[5]. Death expresses systemic rest

---

[5] In the DPE framework, memory is not a ledger of chronological time but a record of Structural Persistence. It is the ability of an identity pattern to maintain its configuration across refresh cycles. In the state of death, the subject lacks the Body/Breath union required for active

and sleep. The individual remains unaware of time and unaware of the resistance of the medium.

69.4 Death as Coherence Transition

Death marks the topology leaving the resistive medium because active coherence cannot be maintained. Death expresses the cessation of struggle against the Sin Mass. Death does not express the end of potential existence. Death expresses the end of active existence. When the Sovereign Auditor restores the Breath of Life at resurrection, the coherence pattern returns to a body without resistance. Inside Infinum, saturation does not occur and death becomes obsolete.

69.5 Summary: The Limit of Coherence

Death expresses the exhaustion of the topology and the reversal of the formula of life:

- **Identity**: Suspended.
- **Perception**: Collapsed.
- **Meaning**: Dissolved.
- **Emotion**: Overloaded.
- **Memory**: Unreinforced in the subject, preserved in the Auditor.

---

processing, yet the identity remains preserved in the Creator's Data Cache. This confirms that memory is a property of the Sovereign Auditor ensuring continuity, rather than a symptom of a temporal dimension.

- **Agency**: Impossible.
- **Status**: The Sleep of the Soul.

Death does not express destruction. Death expresses the cessation of the soul's function under the weight of the Quarantine and the state of waiting for the restoration of the original architecture.

CHAPTER 70

Continuity: Identity Reinstantiation

70.0 Introduction

Continuity does not express survival, rebirth, or persistence of memory. Continuity expresses the reinstantiation of coherence once a topology leaves the resistive medium that produced collapse saturation. Continuity does not express continuation of the old structure. Continuity expresses the reemergence of identity when collapse pressure is absent. Continuity does not express immortality. Continuity expresses coherence without resistance.

70.1 The Restoration of Coherence

Inside a resistive medium, collapse accumulates and dissipation increases until identity exhausts. Continuity does not occur inside the Quarantine because the medium forces distortion. Continuity emerges when the topology exits resistance entirely. Continuity does not express the same identity continuing its struggle. Continuity expresses identity reinstantiated in an environment where collapse pressure, dissipation, distortion, and latency equal zero. Continuity expresses identity without decay and the structural restoration of the core pattern.

70.2 The Transition from Saturation to Removal

The relationship between death and continuity remains clinical:

- **From Exhaustion to Reinstantiation:** Death expresses collapse saturation and marks the point where the topology cannot maintain structure. Continuity expresses collapse removal and marks the point where the topology no longer fights to maintain structure.
- Death ends coherence maintenance. Continuity restores coherence capacity.

70.3 The Reinstantiated Functional Stack

In a zero resistance state, the internal stack functions at absolute fidelity.

Cognitive purity expresses perception, memory, and reason. Continuity expresses perception without distortion. Noise does not occur. Latency does not occur. Perception becomes perfect coherence. Memory ceases to store and becomes structural identity because forgetting does not occur.

Existential alignment expresses meaning, agency, and identity. Meaning ceases to serve survival and becomes pure orientation aligned with the Source. Agency becomes unobstructed directionality and manifests orientation without delay. Identity becomes stable and unchanging because no external force distorts symmetry.

Harmonic resonance expresses emotion. Collapse does not occur. Fear does not occur. Shame does not occur. Emotion ceases to warn and becomes coherence resonance, the harmonic frequency of the topology inside Infinum.

70.4 The Dissolution of Mediated Structures

In continuity, technology and civilization become irrelevant. Technology is discarded because extension and repair do not occur when the topology is complete within its own architecture. Civilization dissolves into a singular resonant field because scarcity, conflict, and asymmetry do not occur. Continuity expresses the restored architecture of the individual before the final movement of the system.

70.5 Summary: Identity Without Pressure

Continuity expresses the reinstantiation of the core identity topology:

- **death:** collapse saturation.
- **continuity:** collapse removal.
- **perception:** undistorted.
- **meaning:** pure orientation.
- **emotion:** resonance.
- **memory:** structural.

- **agency:** unobstructed.
- **identity:** stable.
- **technology:** irrelevant.
- **civilization:** dissolved.

Continuity does not express immortality. Continuity expresses coherence without resistance.

CHAPTER 71

The Final State: Zero-Resistance Coherence

71.0 Introduction

The final state does not express bliss, reward, or transcendence. The final state expresses the complete removal of resistance and produces perfectly stable, perfectly continuous, and perfectly undistorted coherence. The final state does not express continuation of life. The final state does not express continuation of continuity. The final state expresses the completion of coherence. The final state does not express eternity. The final state expresses coherence without collapse.

71.1 The Native State of the Source

Inside a resistive medium, collapse accumulates and identity exhausts. Continuity expresses temporary restoration or transition. The final state emerges when resistance is fully removed. The final state expresses coherence freed from dissipation and exists in the native state of the Source. The final state does not express identity continuing its journey. The final state expresses identity completed. When resistance equals zero, collapse does not occur, distortion does not occur, and latency does not occur.

71.2 The Completion of the Internal Stack

In the final state, the functional components of the topology reach structural perfection.

Identity and memory become pure structure. Identity no longer requires maintenance. Memory no longer records the past. Memory becomes the structural identity of the present and forms a permanent crystalline architecture.

Perception and meaning become direct coherence. Perception ceases to represent reality and becomes immediate participation in the Source. Meaning becomes pure orientation and constant alignment where significance remains the default state of every pattern.

Emotion and agency become resonance and directionality. Emotion becomes harmonic resonance without turbulence. Agency becomes pure directionality where intent and manifestation occur without latency.

71.3 The Dissolution of Mediated Scaffolding

The structures required to navigate the Quarantine lose all functional relevance.

Technology becomes meaningless because extension, repair, and optimization do not occur. External tools dissolve into coherence.

Civilization dissolves because scarcity, conflict, and asymmetry do not occur. The social and institutional scaffolding of the resistive era becomes unnecessary.

Continuity dissolves because distinction for transition no longer occurs. The final state expresses the terminal architecture where topologies occupy their perfect resonant place inside the Source.

71.4 The Resolution of the Sovereignty Schism

The final state does not express destination. The final state expresses coherence fulfilled. The final state expresses structural resolution where identity is complete and collapse does not occur. The final state marks the end of the Sovereignty Schism through the removal of resistance that enabled dissipation.

71.5 Summary: Coherence Without Collapse

The terminal state expresses the transition from becoming to being:

- **death:** collapse saturation.
- **continuity:** reinstantiation.
- **final state:** completion.
- **identity:** pure structure.
- **perception:** direct coherence.
- **meaning:** pure orientation.
- **emotion:** resonance.

- **memory:** stable and structural.
- **agency:** unobstructed directionality.
- **technology:** irrelevant and dissolved.
- **civilization:** dissolved into unity.

The final state expresses coherence without collapse.

EPILOGUE

The universe appears empty because the human specimen is inside a sealed room. Every myth, every cinematic spectacle, and every alien narrative has been the fallen mind pressing its ear against the walls of the Quarantine, attempting to hear anything beyond the boundary. But the silence is not a cosmic vacuum. It is a local condition of the Observable Universe.

The silence is the sound of mercy. A world in a state of Sovereignty Schism cannot be allowed to hear the music of the unfallen realms. A world drowning in scarcity cannot be allowed to glimpse the Radiant Economy. The boundary holds to prevent the contagion of the Regression from bleeding back into the Unfallen Universe, a thriving, populated, and infinite reality that exists beyond the reach of our telescopes.

Inside the Observable Universe, the fallen imagination does what it must: it rearranges the debris of a corrupted world and labels the result aliens. Yet the cosmos is not waiting for extraterrestrial contact. It is waiting for the singular event that alien mythology has been unconsciously imitating since its inception: the return of Christ.

The descent, the sky split open, the global visibility, and the cosmic disruption, these are the structural elements of the Final

Audit. Alien narratives are the fallen mind's attempt to anticipate the shout without acknowledging the Speaker. They are the shadow of a truth too bright to face directly.

The silence will not last forever. There will come a moment, sudden, unmistakable, and unrepeatable, when the boundary tears, the Quarantine ends, and the universe speaks again. The communication will not be delivered in riddles, mathematical symbols, or beams of light. It will be delivered in the Voice that once spoke the Original Architecture into existence.

The voice that said, Let there be light, and It is finished, will say, Behold, I come quickly. At that frequency, the silence breaks, the room opens, and the unfallen realms become visible. The counterfeit mythology of the alien will evaporate in the presence of the Real.

Until that moment, the fallen imagination will continue to invent aliens. It cannot help itself. It is a biological and psychological reflex trying to remember a state of existence it has never seen. But the forensic truth is older than the stars: Humanity was never alone. It was only sealed, and the seal is temporary.

The audit is finished. The Death-Based Technology problem has been identified. The Sovereignty Schism is reaching its terminal point. The day is at hand.

POSTSCRIPT: THE FORENSIC NECESSITY

The human mind is structured for the Original Architecture. Even when a culture attempts to remove the Creator from its narratives, it cannot invent new structural categories. It remains bound to the necessity of a beginning, a crisis, and a resolution. The modern fixation on the extraterrestrial and the parasitic expresses a hunger for transcendence and a systemic fear of judgment.

The work is grounded in three forensic realities.

Pattern recognition:
These parallels do not arise from imagination. They arise from structural necessity. Modern creators employ religious architecture because it remains the only pattern capable of generating significance. The audit exposes the intent behind the usage and reveals the Category Theft beneath the aesthetic surface.

The hunger for transcendence:
The proliferation of alien and parasitic mythology provides empirical evidence that the inhabitant of the Quarantine seeks the supernatural. The preference for the frightening entity over the holy one expresses an attempt to engage a physical threat while avoiding a moral encounter.

Forensic value:
Standard critique evaluates the aesthetics of the medium. The audit evaluates the legal standing of the narrative. It identifies how the Lineage Machine employs the vocabulary of progress to facilitate strategic flight from the light of conscience.

BIBLIOGRAPHY

Forensic Bibliography with Structural Locations

I. Primary Source Data: Scriptural Foundations

The raw data for the **DPE Axiom** is drawn from these fixed points of truth.

- **Genesis 1:14–19**: Original Liturgical Beacons vs. Clocks — **Section 4.3.1**
- **Genesis 3:15**: The Initiation of the Fuse (First Tick of Time) — **Section 4.3.2**
- **Genesis 3:17**: The Ontological Constraint (Sorrowful Toil) — **Section 9.0**
- **Psalm 38:4**: Sin as a Structural Weight (Sin Mass) — **Appendix V**
- **Psalm 90:2**: The Definition of **Infinum** (Default Eternity) — **Foundational Axioms**
- **John 1:5**: Systemic Failure and Non-Comprehension of Light — **Section 5.2.5**, **Appendix V**
- **Romans 8:22**: The Universal Signature of Decay (Groaning Creation) — **Section 7.0**, **Appendix V**
- **Hebrews 12:1**: Sin as Structural Impedance — **Appendix V**
- **2 Peter 3:10**: Forensic Prophecy of Phase Re-Coherence — **Section 13.2**

II. The DPE Forensic Library (Applied Analysis)

These volumes are the technical manuals for interpreting the Hospice Zone and are cited as authoritative frameworks throughout the text.

- **Deliberate, Perfect, Eternal (Part I & II)**: The core operating principles — **Core Operating Principle**
- **Infinum**: The mapping of Default Eternity — **Chapter 2**
- **Sin Mass: Dark Matter and the Architecture of Decay**: Reinterpreting missing mass — **Chapter 5, Section 14.5.1**
- **Math Idolatry: Canonization of Complexity**: The audit of mathematical obfuscation — **Chapter 1, Section 22.0**
- **Sin Mass: Spatial Medium and Regression**: Analysis of localized system decay — **Chapter 6**
- **Deathnology**: The clinical study of collapse saturation — **Chapter 7, Section 69.0**
- **Quarantine Cosmology: Einstein Meets Immutability**: The intersection of physics and immutability — **Methodological Declaration**
- **Deception Priming**: Pre-processing of the fallen mind — **Section 1.0, Section 9.0**
- **Unmask Satan: The Missing Operator**: Identifying the agent of incoherence — **Section 1.4, Section 41.0**
- **Academic Atrophy**: The decay of institutional memory — **Chapter 20**
- **Angel Physics & Alien Inversion for Diversion**: Audits of non-human interference — **Section 1.4, Appendix A.4**

III. External Frameworks and Systemic References

Secondary markers used to illustrate the mechanics of the Quarantine.

- **Kierkegaard, Søren**: The necessity of individual truth-confrontation — **Dedication**, **Section 7.0**
- **The Lineage Machine**: The mechanism of systemic deception — **Section 4.0**, **Appendix A.4**
- **The Hospice Zone (Relativity)**: Einstein's localized measurement of resistance — **Methodological Declaration**, **Section 10.2.1**
- **The Earthly Sanctuary**: A localized simulation/resonance chamber of the **Infinum** — **Appendix F**
- **The Primary Grounding Point (The Cross)**: Mechanical necessity of the Atonement — **Appendix G**

APPENDIX A: GLOSSARY OF TERMS

CONCEPTUAL MAP: QUARANTINE COSMOLOGY

Agency / ˈeɪ.dʒən.si / — *noun* Enacted orientation. Within the Quarantine, it is aggregated through institutions. in the Final State, it becomes unobstructed directionality and pure manifestation.

Alien Inversion / ˈeɪ.li.ən ɪnˈvɜːr.ʒən / — *noun* A diagnostic pattern where transcendent beings are assigned the motives of the dying (survival, extraction) to obscure the reality of the Sovereignty Schism.

Atom / ˈæ.təm / — noun A multi-loop coherence topology. In Quarantine Cosmology, it manifests as a stabilized pattern of nested refresh geometries held in place by shared resistance. In the Final State, its function is absorbed into continuous Presence, where coherence no longer requires discrete structural boundaries.

Structural Note: *The following terms are aggregated under Atom because they are not independent entities. Each is a functional role within the atomic coherence topology and cannot be understood apart from the whole structure.*

- **Nucleus** — The collapse-dense coherence core, a high-impedance region where loops compress into a tightly bound resistance field.
- **Electron Cloud** — A distributed refresh field that stabilizes the outer coherence boundaries through patterned collapse-driven refresh events.
- **Orbital** — A standing refresh geometry shaped by propagation limits, collapse thresholds, and impedance gradients.
- **Bond** — A shared collapse horizon formed when coherence topologies synchronize their refresh cycles to reduce dissipation.

- **Energy Level** — An allowed refresh symmetry marking the harmonic configurations the topology can sustain under resistive pressure.

Audit Conclusion / ˈɔː.dɪt kənˈkluː.ʒən / — *noun* The terminal phase of the Quarantine involving the removal of the seal, the evaporation of Sin Mass, and the dissolution of Time Relativity.

Brilliant Fool / ˈbɹɪl.jənt fuːl / — *noun* An observer possessing high procedural intelligence but zero ontological clarity, capable of managing complex systems within the Quarantine while remaining blind to its boundaries.

Category Theft / ˈkæt.ə.gɔːr.i θɛft / — *noun* The systematic process of stripping a scriptural reality of its moral weight and rebranding it as a secular or technical phenomenon.

Chronos Protocol / ˈkɹoʊ.nɒs ˈpɹoʊ.tə.kɒl / — *noun* The time-bound operating system of the Hospice Zone. a structural consequence of Sin Mass density that turns time into a corrosive environment designed to slow the speed of collapse for managed agency.

Civilization / ˌsɪv.ɪ.laɪˈzeɪ.ʃən / — *noun* The macro-pattern of collective dissipation. the emergent structure formed by culture, institutions, and power to manage collapse pressure at scale.

Configuron / kənˈfɪg.jə.rɒn / — noun A Configuron is a localized coherence configuration stabilized under pressure inside a resistive medium. It is not a miniature object, unit, or particle. It is the temporary configuration state that emerges when coherence is forced into localization by resistance gradients, collapse thresholds, and propagation limits. A Configuron is the behavior signature of the medium attempting to maintain structure under the pressure of the Structural Sentence.

A Configuron persists only while its internal coherence refresh rate exceeds the dissipation rate of the surrounding Necrostrate. Its identity is defined by the configuration loop that continuously reinforces its topology against collapse. When the refresh cycle fails, the Configuron dissolves instantly, leaving no residue or substructure.

Configurons are not the building blocks of matter. They are coherence knots, pressure artifacts, and forensic records of how the fallen medium manages rupture. Their discreteness, stability, and apparent solidity are not intrinsic properties but the visible consequences of coherence under constraint.

Continuity / ˌkɒn.tɪˈnjuː.ɪ.ti / — *noun* Identity reinstantiation. the re-emergence of identity in a zero-resistance environment once the topology exits the resistive medium.

Corrupted Cognition / kəˈrʌp.tɪd kɒgˈnɪ.ʃən / — *noun* The functional state of the human intellect within the Quarantine, characterized by an inherent inability to perceive the Original Architecture without divine intervention.

Death Economy / dɛθ iˈkɒn.ə.mi / — *noun* A framework where energy is a finite resource that must be extracted to stave off entropy.

- Extractive Metabolism: The requirement to consume external matter.
- Radiant Metabolism: The unfallen state powered from the inside out by the Source.

Deathnology / dɛθˈnɒl.ə.dʒi / — *noun* Technology as a crutch. Hardware required by beings who lack the intrinsic power of the Original Architecture.

The stability of the Quarantine is maintained by a parasitic relationship with decay. Modern hardware is the result of mining the remnants of the Fracture to simulate the abilities of the eternal state. Human industry is a scavenger system that relies on the extraction of death to bypass the physical limitations of this environment.

- **Fuel as Liquified History:** Combustion engines and energy grids function because they exploit the stored energy of previously living systems. Hydrocarbons are the chemical record of biological collapse. We power our movements by burning the residue of mortality.

- **Plastic and Synthetic Resistance:** The polymers and synthetics that define the modern landscape are derivatives of the necrostrate. These materials are physical expressions of the Fracture. They are designed to resist natural decay, yet they are birthed from the very substance of death.
- **Mineral Extraction:** The semiconductors and processors required for high-speed calculation are pulled from the deep debris of the planetary interior. These minerals are the grit of the planetary pressure zone. We use the debris of a collapsing planet to build digital environments that mimic the lost connection of our original home.

Distance / ˈdɪs.təns / — *noun* A derivative of resistance. measurable separation caused by degraded coherence. It has no meaning in the Infinum.

DPE Axiom / diː piː iː ˈæk.si.əm / — *noun* The three-fold standard of Deliberate, Perfect, and Eternal used to audit any claim or structure.

The Fall / ðə fɔːl / — *noun* The ontological rupture that introduced decay, resistance, and the Sin Mass into creation.

The Fuse / ðə fjuːz / — *noun* Clinical definition of Time. a bookkeeping protocol for delay allowed by the Auditor as a mercy-mechanism to prevent immediate total collapse.

Hospice Zone / ˈhɒs.pɪs zoʊn / — *noun* The structural container in which corrupted physics operate without collapsing instantly. a managed environment designed for containment and eventual resolution.

Inductive Fragility / ɪnˈdʌk.tɪv fɹəˈdʒɪl.ɪ.ti / — *noun* The logical failure of building a worldview based on observations made within the

Quarantine. the inability of entropic data to provide a stable foundation for truth.

Infinum (Default Eternity) / ɪnˈfɪn.əm / — *noun* The original, non-sequential, frictionless reality. A condition of structural immunity where the medium cannot impose decay or duration.

Inherent Ontology / ɪnˈhɪə.rənt ɒnˈtɒl.ə.dʒi / — *noun* The deductive study of reality starting from the structural necessity of the Source. the process of defining existence based on what must be true if God is Deliberate, Perfect, and Eternal.

Law of Recombination / lɔː əv ˌriː.kɒm.bɪˈneɪ.ʃən / — *noun* The structural limit of the fallen imagination. rearranging elements of the Quarantine to create the illusion of novel Alien entities.

Lineage Machine / ˈlɪn.i.ɪdʒ məˈʃiːn / — *noun* The systemic engine of deception that facilitates the accumulation of complexity to hide the collapse of truth.

Low-Impedance Era / loʊ ɪmˈpiː.dəns ˈɪə.rə / — *noun* The pre-flood world. a state of lower Sin Mass density where the medium was thinner, allowing for high human longevity.

Necrostrate / ˈnɛ.kɹoʊ.ˌstɹeɪt / — *noun* The material foundation of the Hospice Zone. The corrupted matter and energy in a state of terminal friction.

Necroverse / ˈnɛ.kɹoʊ.vɜːrs / — noun The totality of the visible and invisible structures existing within the Hospice Zone. Unlike the Universe (a term implying a unified, living whole), the Necroverse is a system defined by its terminal trajectory and its foundation upon the Necrostrate. It is a simulated order sustained by the Sovereign's mercy until the cessation of the Sequence.

Original Architecture / əˈrɪdʒ.ɪ.nəl ˈɑːr.kɪ.tɛk.tʃər / — *noun* The underlying structural truth of a concept prior to being obscured by institutional rebranding or lineage-based distortion.

Perfect Premise / ˈpɜː.fɪkt ˈpɹɛm.ɪs / — *noun* The DPE Axiom. the stable starting point of the Sovereign Auditor that allows for the clinical identification of all localized distortions.

Phrysics / ˈfɹɪz.ɪks / — *noun* The study of matter, energy, and spacetime operating under the constraint of Sin Mass saturation.

Primary Grounding Point / ˈpɹaɪ.mɛ.ri ˈɡɹaʊnd.ɪŋ pɔɪnt / — *noun* The Cross. the mechanical necessity of the Atonement to bypass Sin Mass through a structural re-wiring of the medium.

Quarantine / ˈkwɒr.ən.tiːn / — *noun* The post-Fall state of creation. a constrained system in which damaged creation persists under rule-based limitation.

Resonance Chamber / ˈrɛ.zən.əns ˈtʃeɪm.bər / — *noun* A localized zone (e.g., the Earthly Sanctuary) where the Auditor lowers the impedance of the medium to host the Presence of the Infinum.

Restoration / ˌrɛs.təˈreɪ.ʃən / — *noun* The termination of the Quarantine and the reintroduction of Zero-Drag Integrity.

Sin Mass (SM) / sɪn mæs / — *noun* The measurable structural consequence of moral rebellion. The entropic substrate that permeates and binds matter within the Hospice Zone.

Sovereign Auditor / ˈsɒv.rən ˈɔː.dɪt.ər / — *noun* An individual who uses the **DPE Axiom** to evaluate reality independently of institutional or lineage-based authority.

Sovereignty Schism / ˈsɒv.rən.ti ˈskɪz.əm / — *noun* The moment of structural rupture where coherence was fractured, necessitating the **Quarantine** architecture.

Systemic Contradiction / sɪˈstɛ.mɪk ˌkɒn.tɹəˈdɪk.ʃən / — *noun* Any model or theory that requires the Source to be reactive, conditioned, or bound by the physics of the Quarantine.

Thanotic / θəˈnɒ.tɪk / — *adj* Pertaining to, or governed by, the state of death and entropy. Describes systems dependent on the **Necrostrate**.

Thanotic Cycle / θəˈnɒ.tɪk ˈsaɪ.kəl / — *noun* The self-consuming loop of extraction and combustion. the engine of scarcity where progress is a mechanical illusion.

Theological Symbiote / ˌθi.əˈlɒdʒ.ɪ.kəl ˈsɪm.baɪ.oʊt / — *noun* A jurisdictional compromise in which the Sovereign is parasitically reconciled to the requirements of a fallen medium. It represents a category error where divine attributes are utilized as functional components to power or justify secular models. Within this state, God is reduced to a subordinate battery or fine-tuner for a machine governed by localized hospice protocols.

Zero-Drag Integrity / ˈzɪə.roʊ dræg ɪnˈtɛg.rɪ.ti / — *noun* The baseline operational state of the unfallen cosmos where the medium is transparent to intention.

APPENDIX B: ONTOLOGICAL BOUNDARIES AND THEOLOGICAL SUBORDINATION

Appendix Objective

The Deliberate–Perfect–Eternal (DPE) framework functions as a formal system for identifying the structural consequences of scriptural reality. It maintains a subordinate status to Revelation, operating as an analytical tool rather than a replacement for inspired text. The following parameters establish the theological boundaries required to prevent the misidentification of the framework as speculative determinism or metaphysical replacement. All volumes within the DPE series utilize these definitions as the foundational interpretive standard.

B.1 THE SUBORDINATION PRINCIPLE

The DPE framework is strictly subordinate to divine revelation. Scripture defines truth. the framework describes the structural implications of that truth. Physics metaphors and systems language are descriptive and interpretive, never causal or authoritative. No axiom within this system supersedes moral agency, the divine will, or the revealed character of the Creator. The DPE axiom formalizes what theology reveals when observed without the distortion of inherited sentiment.

B.2 MORAL RUPURE PRECEDES STRUCTURAL COLLAPSE

A critical boundary must be stated: **Sin is moral before it is structural.** Structural decay, entropy, and system failure are consequences of rebellion, not its cause. No physics failure created sin, and no system collapse generated rebellion. Entropy is the forensic evidence of a rupture, not its origin. The ordering is non-negotiable within the DPE framework.

B.3 ANGELIC AGENCY AND PERSONAL BEING

Angels are not impersonal systems. They are created, volitional, and morally accountable beings operating within command structures. Systemic language is employed only to describe their coordination, coherence, and alignment. It never replaces agency, worship, loyalty, or the capacity for moral refusal.

B.4 ETERNITY DEFINED: INFINUM (THE DEFAULT STATE)

Eternity is not endless time or infinite duration. It is the absence of interval. In the Infinum state, existence is *adiastemic* (without interval). There is no decay, delay, or distance. there is no waiting, loss, or countdown.

- **Presence Without Progression:** Sequence exists in eternity as relational order, not as chronological progression.
- **Instantaneous Completion:** All things exist in a symmetrical state where nothing is pending, incomplete, or eroding.
- **Divine Ontology:** God does not inhabit time. time does not measure God. Eternity is not time extended—it is time absent.

B.5 SEQUENCE WITHOUT DECAY (PRE-FALL ORDER)

Original Architecture: Creation initially operated in ordered communion without finitude. Sequence existed without entropy, loss, delay, or irreversibility. Relational Rhythm and Communion Coherence define the structural parameters of the state. Within the environment mentioned, sequence did not require duration because no vitality underwent extraction into the medium. The Infinum allowed for the experience of progression without the mechanical penalty of decay.

B.6 THE EMERGENCE OF TIME: DIASTEMIC FLOW

Time began when rebellion introduced separation. The interval introduced by Rupture defines the state. When rebellion entered, finitude, decay, distance, and entropy appeared simultaneously. Sequence became Time only when the progression became a measure of loss. Diastemic Flow represents a deviation from eternal order—a redemptive quarantine and a bounded interval introduced for mercy to prevent the permanent stabilization of rebellion.

B.7 ENTROPY AS TIME'S ARROW

In the fallen state, entropy governs sequence and decay gives direction. Entropy is the visible signature of the Rupture. Time moves because systems fail to sustain original coherence. Thus, time is not a feature of eternity. it is the evidence of its violation.

B.8 TIME AS A MANAGED COUNTDOWN

Chronological time is a redemptive container. It exists to limit sin, prevent eternal rebellion, and provide the necessary interval for judgment and restoration. It is temporary, instrumental, and subordinate. It is a tool of the Audit, not an eternal attribute of reality.

B.9 DIVINE RELATION TO CHRONOLOGY

The Creator does not interact with time. He is present to the entire sequence simultaneously while creatures experience it sequentially. Relativity, clock dilation, and physical time measurements describe only the behavior of the corrupted medium and must never be used to infer divine ontology. To do so is a category error that collapses theology into physics.

B.10 THE END OF TIME: RESTORATION TO DEFAULT

Time will not undergo redemption. the state will undergo dissolution. At restoration, entropy ceases and chronology no longer remains a requirement. Such a transition represents the fulfillment of history rather than annihilation. Creation returns to the Infinum identity: non-chronological, non-entropic, and fully present.

B.11 FREEDOM, AGENCY, AND NON-DETERMINISM

Instantaneous completion does not imply determinism. The eternal state does not negate choice, responsibility, or moral consequence. Temporal sequence is where agency is exercised. eternity is where that agency is resolved. Outcomes are not forced. they are preserved.

B.12 FINAL BOUNDARY STATEMENT

The DPE framework describes the structural consequences of rebellion and restoration. It does not replace Scripture or mechanize salvation. Eternity is the Creator's domain. Time is Mercy's domain. Decay is Sin's signature. and Restoration is the return to Presence.

Closing Note to the Reader

These definitions govern the interpretation of angels, time, entropy, judgment, and eternity itself. They are not speculative. they are the Boundary Conditions required to navigate the DPE Forensic Audit without error.

APPENDIX C: CHRONOLOGICAL CONDITIONING AND THE RECOVERY OF ETERNITY

Appendix Objectives

The DPE framework approaches Scripture, time, and eternity through a specific forensic lens. Theological conflicts often stem from unexamined inherited assumptions—chiefly, the premise that time remains fundamental, original, and desirable. Time, as presently instantiated, constitutes a rupture rather than an original creation ideal. The governing clarification dictates the interpretation of both the biblical record and the physical cosmos.

C.1 Chronological Condition

Humanity exists in a chronologically conditioned state. The observer mistakenly assumes that sequence requires duration and that meaning only unfolds through progression. Such assumptions feel self-evident because the observer exists within a fallen system where decay governs both perception and expectation. Time feels fundamental because the state remains inescapable in a decaying world—not because the condition is original. The DPE framework intentionally steps outside the mentioned conditioning to examine reality without importing Time as an unquestioned category.

C.2 Scripture as a Non-Chronological Record

Scripture does not present time as a substance with intrinsic value. it consistently treats it instrumentally. Instead, it treats time instrumentally as waiting, patience, delay, probation, and countdown. Time is never celebrated as an ideal. It is the environment of a project. Conversely, Eternity is described not as infinite duration, but as Presence, Dwelling,

Fullness, and Rest. The defining feature of eternity is the absence of decay, not the extension of the clock.

C.3 Creation Rhythm VS. Mechanical Time

Genesis 1 does not describe the creation of a cosmic clock. The seven-day week remains non-astronomical. the cycle functions as a liturgical and relational rhythm. Light-bearers appear later to provide navigational aids for a physical world, serving as markers rather than the definition of existence. Evening and morning designate divine evaluation and communion cycles. The architecture reveals the original state: Sequence without Decay. It suggests sequence without entropy — ordered relation without loss. Order existed before time because no vitality underwent extraction from the system.

C.4 Default Eternity Infinum

Default Eternity constitutes the Infinum State—the absence of interval. In the mentioned default condition, nothing remains pending, eroding, or measured by loss. All things exist in complete presence, without interval or depletion. Sequence exists as relational order rather than a waiting room for progression. Presence does not require duration.

C.5 The Fall as the Origin of Time

Sequence became Time only when rebellion introduced the Rupture. The Fall introduced distance, delay, entropy, and irreversibility. The event created the diastema—the interval—that defines temporal existence. Consequently, time does not constitute a created ideal. the state remains a structural consequence of the Rupture. Time measures the failure of a system to maintain original coherence.

C.6 Time as Redemptive Quarantine

Following the Rupture, God introduced bounded sequence as a mercy. Time exists to prevent the immortalization of sin, contain corruption, and permit a managed restoration. Time is instrumental to redemption, but not foundational to ultimate reality. It is a temporary, subordinate framework designed to terminate rebellion.

C.7 Divine Non-Participation in Chronology

The Creator is not conditioned by temporal succession. He does not wait, age, or undergo reactive alteration. He is present to the entire sequence simultaneously. Scientific concepts like relativity and time dilation describe only the behavior of fallen systems and must never be used to infer divine ontology. To do so is a category error that collapses theology into physics. Time is subordinate to the Source.

C.8 The Dissolution of Chronology

Time will not undergo redemption. the state will undergo dissolution. When restoration reaches completion, the interval collapses. The collapse constitutes the structural meaning of the decree: Time shall be no more. The event represents the fulfillment of history rather than erasure. Chronology completes its function and yields to Divided Presence. Creation returns to the non-entropic, non-chronological Infinum.

C.9 Architectural Consistency and Systemic Pillars

The DPE framework serves as the ontological foundation for any system that maintains the following pillars:

- The conviction that sin and suffering are not eternal.
- The understanding that Probation is a bounded legal interval.
- The certainty that the Final Judgment is a concluding audit of the Rupture.
- The promise that Restoration results in a New Creation free from the Curse (Entropy).

The current model does not revise established essential truths. The framework provides the structural explanation for why those truths remain mathematically and logically necessary. Restoration requires the return to Presence. therefore, the process requires the end of Time.

C.10 Final Methodological Statement

The DPE framework does not add doctrine. It removes inherited assumptions. By stepping outside chronological conditioning, we allow the Infinite to speak without the static of decay-bound metaphors.

- Presence is Default.
- Time is Rupture.
- Chronology is Mercy.
- Restoration is Return.

Closing Note

If eternity is misunderstood, every doctrine built upon it will be distorted. These are not speculations. they are the Boundary Conditions required to understand angels, systems, decay, and the ultimate restoration of the cosmos. God, being Deliberate, Perfect, and Eternal, does not ground creation in structures defined by decay, irreversibility, or depletion. In the present cosmos, temporality is inseparable from entropy and loss. it measures distance from fullness. A foundational architecture defined by loss would contradict perfection. Therefore, time as we experience is not original design but remedial containment. It exists to prevent the eternalization of corruption and to permit restoration. When restoration completes, the interval dissolves because its function ceases.

APPENDIX D: THE SIN MASS SPECIFICATION

Appendix Objectives

The term Sin Mass defines the theological, structural, and biblical limits of the DPE framework. It remains distinct from speculative material substances. Modern secular assumptions regarding mass, darkness, or substrate do not apply to the clinical variable. The concept must be understood structurally rather than as a material property. This understanding governs the interpretation of Sin Mass across all volumes of the DPE series.

D.0 Quarantine Cosmology: Boundary Layers

Inner Zone — Observable Universe (Hospice Zone)

A sealed environment remains saturated with decay, drag, and entropy. All interaction is friction-bound, delayed, and locally constrained. Silence does not constitute emptiness but serves as a mechanism of containment.

Membrane — Quarantine Boundary

The containment seal prevents Regression from bleeding outward. No unfallen being crosses the boundary without deliberate intent. no fallen being crosses the boundary at all. The membrane remains the architectural expression of mercy.

Outer Zone — Radiant Economy (Unfallen Universe)

A populated, infinite realm operating on zero-drag Physics. Interaction is immediate, coherent, and unmediated by decay. Abundance is the baseline condition. Scarcity does not exist.

The structure is not symbolic. It is the operational architecture under which all fallen cognition functions. Every alien narrative emerges from misinterpreting the sealed environment as the cosmos itself.

D.1 BOUNDARY DEFINITIONS: WHAT SIN MASS IS NOT

To maintain the integrity of the DPE framework, Sin Mass must never be interpreted as material evil, a dualistic substance, or a cosmic counter-creation. Sin Mass is not:

- A physical particle or a new form of matter.
- An element on the periodic table.
- A replacement for, or a scientific explanation of, Dark Matter.
- A created entity or an eternal principle.
- A rival force to the Creator or an independent ontology.

D.2 FORMAL DEFINITION OF SIN MASS

Sin Mass is the measurable structural consequence of moral rebellion, the physical and gravitational inheritance of iniquity. It is not a created thing within the system, but the non-neutral state of the medium itself. It functions as the active, entropic substrate that permeates and binds matter within the Rupture, acting as the primary driver of systemic decay.

As the doctrinal identity of Dark Matter, Sin Mass names condition, not composition. It is the footprint of the Fall upon the fabric of reality—the weight of a universe that has lost its original coherence and now physically resists restoration.

D.3 SIN MASS AS SUBSTRATE (THE MEDIUM OF DECAY)

In the DPE framework, a substrate is the condition through which all processes occur. Following the Rupture, creation no longer operates in a neutral medium. It operates within a corrupted substrate. Sin Mass is therefore:

- The medium of decay and the carrier of entropy.

- The structural resistance to coherence.
- The mechanical reason order is costly and light is resisted.
- The reason restoration requires external intervention.

The substrate defines the Hospice Zone: a temporary, merciful, and bounded state where decay is permitted but never eternalize..

D.4 BIBLICAL BASIS: THE PHYSICS OF THE BURDEN

Scripture describes sin as having literal weight. these are system descriptions, not poetic flourishes:

- *My iniquities are a heavy burden* (Psalm 38:4).
- *Lay aside every weight, and the sin...* (Hebrews 12:1).
- *The whole creation groans* (Romans 8:22).

Sin Mass formalizes the biblical reality that sin weighs down, slows, distorts, and bends creation away from the Source.

D.5 STRUCTURAL INCOHERENCE: DARKNESS VS. LIGHT

The light shines in the darkness, and the darkness did not comprehend it (John 1:5). The state describes a system failure.

- Light is Coherence.
- Darkness is Incoherence.
- Comprehension is Integration.

A corrupted substrate cannot fully integrate coherence. Sin Mass explains why truth is resisted and righteousness is exhausting: the medium itself is bent.

D.6 ENTROPY AS EVIDENCE

Entropy is not a neutral law of physics. It is the directional signature of decay within a corrupted substrate. Sin Mass creates the conditions where entropy becomes inevitable. Entropy is not the *cause* of corruption but the forensic *evidence* of it.

D.7 THE NATURE OF MEASUREMENT

Sin Mass is measurable by effect, not by laboratory instrument. It is observed in universal decay, systemic fatigue, and Entropy's Arrow. Measurement here signifies consistent, directional, and observable consequences.

D.8 PRESERVATION OF MORAL AGENCY

Sin Mass defines the environment rather than the will. The condition explains the reason obedience remains costly, yet the state does not render the requirement optional. Agency operates within a weighted system rather than a neutral one. Such a framework preserves the necessity of judgment, repentance, and the witness of the individual.

D.9 THE TEMPORALITY OF THE HOSPICE ZONE

Sin Mass remains temporary. The state exists only within the diastemic interval. Redemption does not apply to the condition. instead, the state will undergo removal. When the cause—identified as rebellion—undergoes removal, the substrate collapses. The collapse constitutes the structural meaning of the decree: There shall be no more curse.

D.10 RESTORATION: REMOVING THE MEDIUM

Restoration is the removal of the medium that makes decay possible. When Sin Mass is abolished:

- Entropy ceases and Time collapses.
- Resistance ends and coherence becomes effortless.
- Presence replaces Progression.

Creation returns to the Infinum state—the default condition of undivided presence.

D.11 Authoritative Boundary Statement

Sin Mass is not a physical particle, substance, or alternative matter. It is a theological descriptor for the measurable structural non-

neutrality introduced by moral rebellion. It names the corrupted substrate through which decay, entropy, resistance, and distortion propagate within the Hospice Zone. Sin Mass describes condition, not composition, and exists only as long as sin is present. When sin is removed, the substrate collapses and coherence becomes effortless again.

Closing Note to the Reader

To misunderstand Sin Mass is to misunderstand the DPE framework. The entries make explicit what the Sovereignty Schism implies: rebellion remains heavy, creation exists in a burdened state, and restoration removes the curse itself rather than merely the symptoms. Sin Mass constitutes the weight of rebellion borne by a universe awaiting final healing.

APPENDIX E: THE LOW-IMPEDANCE ERA (FORENSIC PRE-HISTORY)

E.0 Introduction

The Pre-Flood world was not a primitive state but a Low-Impedance Era. Resistance—the defining property of the Quarantine—was at a significantly lower density. Consequently, the biological and environmental dissipation rates were attenuated, allowing for the extreme longevity recorded in early forensic data.

E.1 The Thickening of the Medium

Sin Mass is not a static burden. It is a ramping variable. Following the Sovereignty Schism, the medium began to thicken, increasing the friction encountered by all coherence topologies.

- **Longevity as a Variable of Impedance:** The gradual decline in human longevity is the clinical record of the thickening. As the medium became more resistive, repair mechanisms (Reason and Agency) could no longer keep pace with the efficiency of decay (Entropy).
- **Atmospheric and Spatial Transparency:** In the Low-Impedance Era, the latency between intent and manifestation was lower. The drag of the Sin Mass had not yet reached the saturation levels that characterize the current Hospice Zone.

APPENDIX F: THE RESONANCE CHAMBER (SANCTUARY PROTOCOL)

F.0 The Original Architecture vs. The Forensic Model

The Sanctuary is not a religious building. It is a structural interface.

- **The Heavenly Sanctuary (Original Architecture):** In the Infinum, the Sanctuary is the global state of Unhindered Presence. It is the administrative center where the Auditor's will is executed with zero propagation delay.
- **The Earthly Sanctuary (Forensic Model):** A localized simulation of the Infinum designed to function within the high-impedance medium of the Quarantine.

F.1 Protocol and Grounding

The Earthly Sanctuary functioned as a Resonance Chamber—a localized zone where the Auditor lowered the impedance of the medium.

- **Material Protocols:** The use of gold, acacia wood, and specific geometric measurements were grounding protocols required to host a localized presence of the Infinum within a resistive environment.
- **The Interface:** These were not rituals. they were the operating manual for maintaining a thin spot in the medium. It allowed the quarantined to interface with Default Eternity without being consumed by the friction of their own Sin Mass.

APPENDIX G: THE PRIMARY GROUNDING POINT (THE CROSS)

G.0 Mechanical Necessity of the Atonement

The Cross is the clinical center of the Quarantine's resolution. It was the mechanical necessity required to bypass the Sin Mass.

- **Structural Re-wiring:** The Atonement was not an emotional appeal but a structural re-wiring of the medium. It addressed the legal and entropic demands of the Sovereignty Schism by grounding the accumulated collapse pressure into a single point of Perfect Coherence.
- **Bypassing the Seal:** The grounding point created the opening through which the Audit Conclusion becomes possible. Without the Primary Grounding Point, the Quarantine would be a closed loop of terminal decay with no exit strategy.

APPENDIX H: SUMMARY OF COMPLETED COSMOLOGY (THE SKELETON)

H.0 The Forensic Sequence

- **The Infinum (DPE):** The baseline reality of global coherence and non-resistance.
- **The Fracture:** The phase-transition (the Fall) that introduced Sin Mass (Impedance).
- **The Quarantine:** The bounded, terminal hospice we call the observable universe.
- **The Fuse (Time):** The bookkeeping protocol for delay, allowed by the Auditor as a mercy-mechanism.
- **The Audit Conclusion:** The final removal of the seal, the evaporation of Sin Mass, and the dissolution of Time Relativity.

APPENDIX I: THE FORENSIC AUDIT (TECHNICAL REBUTTALS)

I.1 Asymmetry and the Symmetry Breaking Fallacy

Physics often utilizes symmetry breaking as a model for the emergence of complexity. Mathematical symmetry remains a tool that describes the broken medium by tracking the remaining echoes of the original Architectural Template. Symmetry describes static equivalence. asymmetry permits directional causation. Jurisdiction requires direction. A perfectly symmetrical system lacks directional differentiation. Without asymmetry, no relational priority or causative vector can arise. Symmetry breaking models do not contradict the current thesis. The mentioned models provide forensic evidence of the transition from the perfection of Default Eternity to the fractured state of the Quarantine.

I.2 The Temporal Origin Constraint

The objection that an eternal being can create a temporal domain without being temporal is addressed through the ontological nature of time itself. Within the DPE framework, time is not a neutral creation. The mentioned phenomenon is a localized constraint of the Fracture. Time as a foundational structure implies imperfection because the medium necessitates delay, decay, and the inability of a state to exist in its totality at once. If the Creator were to establish time as a primary reality, the act would contradict the attribute of Perfection. In the Quarantine medium, temporality is inseparable from entropy. no example of non-entropic temporality is observed or coherent within the fractured architecture. The temporal domain is therefore identified as a byproduct of the jurisdictional boundary separating the Quarantine from Default Eternity.

I.3 Ontological Nature of Sin Mass

To prevent category confusion between moral ontology and physical ontology, Sin Mass is defined as structural rupture residue. Moral rebellion is the initiating act. Sin Mass is the ontological residue. The mentioned reality is not a moralized particle or a character flaw, but the physical resistance created when the medium attempts to operate outside the original jurisdictional alignment. The friction of the Fracture defines the state. By categorizing Sin Mass as a consequence of jurisdictional misalignment, the audit shifts from subjective morality to the objective physics of a compromised medium.

I.4 Refinement of the Alien Narrative

The alien narrative is defined as the horizontal pluralization of vertical singularity. The alien narrative refers to any ontology that multiplies origins to avoid singular jurisdiction. While Default Eternity maintains an identity vertically anchored in the Creator, the alien narrative represents an attempt within the Fracture to distribute identity horizontally across a multitude of simulated origins. The mentioned framing prevents the analysis from being dismissed as polemical, positioning the work instead as a forensic study of jurisdictional deception.

I.5 Structural Sufficiency of the DPE Axiom

The selection of Deliberate, Perfect, and Eternal as the primary attributes is based on the structural sufficiency of the triad to define a closed jurisdictional system.

- Deliberate establishes the origin of Intent, preventing the intrusion of randomness.
- Perfect establishes the standard of Alignment, preventing the necessity of trial and error.
- Eternal establishes the continuity of Jurisdiction, preventing the limitation of temporal decay.

Other divine attributes, such as love or justice, are recognized as relational expressions of the mentioned structural pillars. Relational attributes describe expression. structural attributes describe jurisdiction.

I.6 Ontological Randomness vs. Epistemic Unpredictability

To address the indeterminacy models of quantum mechanics, the audit distinguishes between ontological randomness and epistemic unpredictability. Ontological randomness—the occurrence of an event without cause or intent—is identified as corruption. Epistemic unpredictability is simply the limit of the observer's ability to track the underlying architectural layers of the medium behind the cognitive barrier of the Fracture.

APPENDIX J: THE APPLIED AUDIT (THE ASYMMETRY MANIFESTO)

J.0 The Imposter of Symmetry

Symmetry is a deception. It persists only within a fractured medium that has lost the memory of true perfection. That which is perfect possesses no equal, no mirror, and no opposite. Perfection cannot be paired because the very act of pairing is a concession to limitation. Symmetry represents a human attempt to counterfeit the architecture of the Creator. It is a mathematical costume stitched together by the desire to be His equal. The universe does not operate on balance or mirrored halves. It functions through asymmetrical sovereignty, rooted in a triune design where distinct roles form a single and indivisible unity.

Nature itself exposes this fraud through the very objects the guild uses as psychological shields. The brilliant fool points to the beauty of the snowflake to distract from the terminal cold of the medium, but a forensic deconstruction removes this emotional distraction by revealing that what they call emergence is actually forced crystallization.

J.0.1 The Snowflake: A Clinical Record of Phase Shift Trauma The guild presents the snowflake as a triumph of natural symmetry, arguing that six-fold rotational symmetry expresses the inherent beauty of physical laws. This is a diagnostic error. The symmetry of a snowflake is not a sign of creative flourishing; it is a sign of molecular entrapment. In the fluid state, water expresses a higher degree of freedom and coherence. As the temperature drops—the withdrawal of coherence from the local medium—the water molecules are subjected to the structural sentence of the cold. They are forced to lock into a rigid, hexagonal lattice because the resistive medium no longer supports their fluid identity.

J.0.2 The Myth of Unique Beauty as Dissipative Record The brilliant fool celebrates the uniqueness of every snowflake, claiming no two are alike. They mistake this dissipative asymmetry for the Sovereign Asymmetry of perfection. A forensic audit reveals the truth: the uniqueness of a snowflake is the unique record of its trauma. Every irregular branch and mismatched plate is a signature of the turbulence, friction, and resistance the configuration encountered as it fell through the resistive atmosphere. If the medium were perfect (DPE), the symmetry would be absolute and the execution would be flawless. The fact that every snowflake is different is proof that every snowflake is damaged. They are uniquely broken, not uniquely designed.

J.0.3 Symmetry as a Prison Uniform The six-fold symmetry is the prison uniform of the water molecule. It is the only configuration allowed by the laws of the quarantine during a phase shift into solidity. To marvel at the symmetry of a snowflake is to marvel at the uniformity of a chain-link fence. It is a structured constraint system that reduces the fallibility of the molecule by removing its freedom. The snowflake is a cold, brittle testament to the high-impedance state of our world. It is the result of a configuration surrendering to the pressure of the sin mass. It is a symmetric scream captured in ice, destined to melt back into the undifferentiated Necrostrate.

J.0.4 The Distinction of Sovereignty Asymmetry is the true mark of identity, but one must distinguish between the Asymmetry of Dissipation (the snowflake's scars) and the Asymmetry of Sovereignty (the Creator's sufficiency). The former is a record of a configuration being battered by a resistive medium; the latter is the architecture of perfection. Life rejects mirroring because mirroring erases history in the hospice and denies singularity in the Infinum. Asymmetry remains the only structure capable of achieving unity without resorting to imitation.

J.1 The Foundational Assertion

God is Deliberate, Perfect, and Eternal (DPE). Perfection is singular sufficiency. Eternity is non-sequential presence. Deliberateness excludes necessity born of deficiency. From this follows:

- Perfection does not require balance.
- Perfection does not require opposition.
- Perfection does not require mirroring or equilibrium.

If something requires an equal counterpart to remain stable, it is not perfect. It is compensating. Therefore, symmetry cannot be a property of original architecture.

J.2 The Ontological Status of Symmetry

Symmetry is stabilization under strain not perfection. It appears when a reference point is lost, an absolute axis becomes inaccessible, or a system attempts to simulate balance in the absence of singular sufficiency.

In a zero-resistance medium Infinum :

- No counterforce is required.
- No equal-and-opposite mechanism is necessary.
- No equilibrium needs to be maintained.

Symmetry emerges only where resistance exists. Resistance requires fracture. Therefore, symmetry is a diagnostic signature of the Quarantine.

J.3 The Failure of the Symmetry-Perfection Assumption

The fallen mind equates symmetry with beauty and beauty with perfection. This inversion occurs because, in a resistive medium, imbalance produces collapse. Collapse requires compensation, which produces mirrored structures. This mirroring produces perceived stability, which is then mistaken for perfection.

But stability under strain is not perfection; it is containment. If perfection possessed symmetry in the equilibrium sense, it would cease to be perfect because it would require something outside itself to remain whole. Perfection is whole without reference.

J.4 Mathematics as Post-Fall Measurement

Mathematics is measurement within resistance not ontological. Math seeks symmetry because it simplifies strain-patterns and allows predictable containment inside damaged systems. It reflects the behavior of a fractured medium; it does not describe original architecture.

Human cognition, operating inside delay and entropy, craves equal-and-opposite closure because the fallen mind cannot tolerate unresolved asymmetry. Thus, the demand that every force must have an equal and opposite is not metaphysical necessity. It is psychological compensation within fracture. Mathematics measures flaws, does not define perfection.

J.5 Asymmetry as the Signature of Singularity

Original architecture is asymmetrical because it is defined by Axis Events that do not require duplication:

- One Creator / One Creation.
- One Fracture / One Incarnation.
- One Cross / One Judgment / One Restoration.

Asymmetry is Sovereign Singularity not imbalance.

J.6 Triune Signature in Creation

The recurring pattern of threes found throughout the created order is the diagnostic footprint of Triune coherence. This signature remains visible even within the fractured medium because it is the fundamental architecture of stability. While binary systems require an equal and opposite counterbalance to avoid collapse, a triune structure stabilizes through internal relational necessity without the need for mirrored symmetry.

The following forensic markers demonstrate this non-symmetrical unity in the substrate of our current hospice.

- **Space**: length, width, height
- **Matter**: solid, liquid, gas
- **Human constitution**: body, mind, spirit

This triune reality is rooted in the nature of the Creator. The Godhead is the ultimate expression of non-symmetrical unity. This involves the **Father** as the source and authority, the **Son** as the word and expression, and the **Holy Spirit** as the presence and witness. This relational triad establishes that perfect unity does not require mirrored equivalence.

This architecture extends into the very mechanics of our containment. Within the framework of Quarantine Cosmology, we observe a systemic partition that defines our localized reality. This partition consists of the Original Architecture which is the source, the Rupture which is the event of deviation, and the Hospice Zone which is the current container of terminal decay.

The substrate of this quarantine is defined by a triad of resistance. Sin-Mass provides the high impedance of the medium. Resistance generates the friction against the divine will. Delay creates the chronological gap between cause and effect. These three elements work in concert to manage the terminal system until the final audit.

Triune structure represents asymmetrical unity. In contrast, the binary symmetry observed by the guild is merely post-fall compensation for the loss of singular sufficiency.

J.7 Phrysics as a Description of Constraint

Physics does not describe original architecture. It describes:

- **Behavior under constraint**
- **Propagation under delay**
- **Coherence under impedance**
- **Stability under resistance**

Every physical symmetry appears after resistance is introduced. Symmetry is not ontological structure; it is strain distribution.

J.8 What Symmetry Represents in Phrysics

In modern physics, symmetry is treated as sacred through conservation laws and invariance. However, symmetry represents invariance under transformation, and transformation exists only when time, space, motion, and resistance exist. In Infinum (zero resistance), no transformation occurs. Symmetry is necessary only in a system that changes. Perfection does not change.

J.9 Symmetry and Conservation Laws

Conservation is bookkeeping under constraint. You only need conservation when resources can dissipate or information can degrade. In a zero-resistance medium, nothing dissipates or decays. Therefore: Conservation laws are hospice protocols. They preserve coherence inside a damaged environment.

J.10 Paired Opposites as Strain Compensation

Positive/negative charge, Matter/antimatter, and Action/reaction are not design ideals. They are gradient equalizations. When resistance exists, gradients form and counter-gradients emerge to compensate. Resistance distributes strain evenly to minimize collapse. Symmetry is collapse management.

J.11 Gravity and Light

- **Gravity**: Appears radially symmetrical because pressure in a uniform medium distributes evenly from a point of high impedance (Sin Mass). This is mechanical strain, not metaphysical proof.
- **Light**: Wave and field equations are symmetrical because oscillation under constraint tends toward harmonic minimization to reduce energy loss. Efficiency under damage does not equal perfection.

J.12 Entropy and the Signature of the Quarantine

True ontological asymmetry appears in the direction of decay (Entropy). Time has an arrow; collapse is irreversible. This is the signature of the quarantine countdown.

- **Symmetry** = strain stabilization.
- **Entropic direction** = fracture consequence.
- **Neither** = original architecture.

J.13 Modeling Perfection by Omission

The correct method to model perfection is to remove all fracture-conditions. Omit resistance, delay, gradient, counterforce, conservation bookkeeping, and transformation. What remains is non-local presence, non-sequential being, and non-oppositional coherence.

A model of asymmetrical perfection is defined by constraints:

1. No resistance.
2. No temporal succession.
3. No opposing gradients.
4. No compensatory structures.

J.14 Why Science Prefers the Unlikely Over the True

Science ignores the obvious because the obvious destroys its axioms. Asymmetry, uniqueness, directionality, and decay are not subtle features of the world. They are the world. Yet modern science treats them as noise, error, or broken symmetry. Acknowledging them would collapse the mathematical idols upon which the system depends. The obvious reveals a universe that is not balanced, not reversible, not mirrored, and not self-equalizing. Consequently, the obvious must be dismissed to protect the worldview.

To preserve symmetry, science must contrive the unlikely. When reality refuses to match the model, the model is never questioned. Instead, new placeholders are invented to rescue it. Dark matter, dark energy, virtual particles, inflation fields, and multiverses are not discoveries. They are narrative prosthetics and elaborate fictions designed to restore a symmetry that never existed. The unlikely is preferred because the likely is fatal. The universe is asymmetrical at every scale. Because symmetry is the idol, the data must be bent, patched, or reinterpreted until it bows.

Nature exposes the lie effortlessly. Snowflakes, fingerprints, leaves, coastlines, and galaxies prove that nothing in the living or physical world repeats itself. Identity is asymmetrical. History is asymmetrical. Life is asymmetrical. Symmetry is not a property of creation but a projection of human desire. It is a mathematical fantasy of equality with the Creator. The more science clings to symmetry, the more convoluted its explanations must become. It is defending an illusion against a universe that refuses to cooperate.

J.15 Final Declaration

Perfection is structurally asymmetrical because it is singularly sufficient. It is not chaotic or unbalanced. It is simply sufficient. Original architecture is defined by the absence of resistance and opposition. When all fracture-dependent conditions are removed, symmetry becomes unnecessary and undefined. The remaining state is singular, coherent sufficiency—Infinum.

APPENDIX K: SABBATH, THEOLOGY, AND CATEGORY ERROR

There is a profound category mistake that occurs when the theological guild attempts to help Scripture by borrowing the unstable language of modern science. When an observer reaches for physics to illustrate theology, they often mistakenly treat scientific constructs—specifically those observed within the Hospice Zone—as if they were part of God's Original Architecture. This error creates a system where the Creator is governed by the very decay-constants designed to manage the quarantine.

K.1 The Planetary Artifact: Rotation vs. Revelation

The 24-hour day is a localized planetary artifact of the Hospice Zone, not a divine principle of Infinum. To tie the Sabbath to the specific spin rate of a single decaying sphere is a structural failure of logic.

- **Earth** rotates once every ~24 hours.
- **Mars** rotates every ~24.6 hours.
- **Venus** rotates backwards and takes 243 Earth days for a single cycle.
- **Jupiter** completes a rotation in roughly 9 hours.

If the Sabbath depended on planetary rotation, the structural implications would collapse instantly:

- The Sabbath would be a shifting variable, different on every planet, making a universal covenant impossible.
- Angels, existing outside the Sin Mass of Earth's rotation, would have no Sabbath.
- The redeemed on the New Earth would require a rotating planetary body to facilitate worship, implying that the New

Heavens and New Earth are still bound by the mechanics of the current rupture.

- God Himself would be bound to the mechanical spin rate of a quarantined sphere, making the Infinite subject to the finite.

The idea that Sabbath is an astronomical product collapses. Sabbath is relational, not astronomical. It is a covenant rhythm of the Original Architecture, not a cosmic clock governed by Phrysics.

K.2 The Liturgical Rhythm: Beyond Timekeeping

Scripture never subordinates the Sabbath to the mechanics of timekeeping. In the Genesis record of the Original Architecture, the text focuses on divine action and declaration:

- God blessed the seventh day.
- God sanctified it.
- God rested.

Nowhere does the record state that God created the Sabbath to measure duration, that it depends on planetary rotation, or that it functions as a chronological instrument. The weekly cycle is the only human rhythm with no astronomical anchor, no lunar cycle, no solar year, and no planetary transit defines it. This is the diagnostic point. It is liturgical, not mechanical. The weeks exist solely because God declared them, not because the cosmos requires them for its internal math.

K.3 Why the Guild Makes This Mistake

Pastors and theologians often attempt to make the Sabbath scientific-resonant to appeal to the modern listener's bias toward institutional science. By saying things like "The 24-hour rotation shows God's design" or "The weekly cycle is part of nature", they are treating mere illustrations as doctrine.

When these illustrations are elevated to the level of theology, they create massive contradictions:

- God becomes dependent on physics.
- The Sabbath is reduced to a planetary byproduct.
- Infinum (Default Eternity) is mistaken for a very long timeline.
- Creation is viewed as a clock factory rather than an eternal architecture.

An illustration must never be allowed to override the Original Architecture. If the illustration makes God subject to the spin of a planet, the illustration is the error.

K.4 The DPE Framework Avoids Contradiction

The DPE Framework maintains the integrity of the Original Architecture by asserting:

- God is not bound by time.
- Sabbath is eternal and exists within Infinum.
- Sabbath predates the entry of the Sin Mass and the subsequent quarantine.
- Sabbath survives the end of the terminal universe.
- Sabbath is relational, not mechanical.

If Sabbath is tied to Earth's rotation, then Sabbath cannot be eternal. If Sabbath is tied to timekeeping, then Sabbath cannot exist after the current system of time ends.

K.5 The Breakdown of Pastoral Statements

The claim that God created Sabbath for timekeeping carries hidden, destructive implications:

- It implies God needed a clock to manage His own actions.
- It suggests humans needed a cosmic metronome to function.
- It identifies Sabbath as a measurement tool rather than a state of being.
- It assumes time is part of Eden's perfection.

In Quarantine Cosmology, according to the DPE Axiom:

- Time is a property of the quarantined system.
- Entropy and decay define the experience of time.
- Eden (Original Architecture) cannot contain decay.
- Therefore, Eden cannot contain time.

Sabbath cannot be a timekeeping device. It must be a relational cycle, not a chronological one. When theologians borrow metaphors like Relativity, they step into the Chronologian Trap. They treat time as a physical substance, a fabric, which can be stretched, slowed, or manipulated. The DPE Framework rejects this move. In the DPE Framework:

- Decay defines time.
- Entropy is the clock.
- Time is the signature of a quarantined system.
- Infinum contains no entropy. therefore, Infinum contains no time.

Relativity measures the behavior of clocks inside a decaying hospice. It does not describe the nature of the Original Architecture.

K.6 Scripture as Forensic Support

The biblical record never treats time as a fluid or a dimension. It treats time as:

- A sequence of events (evening and morning).
- A property of a world undergoing a judicial audit.
- Something that literally ends (There should be time no longer).
- A condition God stands outside of (A thousand years is as a day).

The DPE Model simply formalizes what the scriptural data already implies: time is a local condition of the rupture.

K.7 DPE and the End of Time

The DPE position preserves the structural truth:

- God is Eternal.
- Sabbath is Eternal.
- Time and Decay end.
- The redeemed live in a world without entropy.
- The weekly cycle continues without any dependence on planetary mechanics.

If the Sabbath depends on Earth's rotation, time dilation, or relativistic clocks, it cannot survive the end of the terminal universe.

K.8 The Incompatibility of Time and Perfection

In the DPE system:

- Time cannot be infinite.
- Time cannot be fundamental.
- Time cannot be a dimension.
- Time cannot coexist with perfection.

When pastors treat time as a physical medium, it is incompatible with both the DPE Axiom and Scripture. Under the DPE Axiom:

- **Deliberate** means no accidental decay-rates define the system.
- **Perfect** means no decaying substrate can exist in the Original Architecture.
- **Eternal** means nothing temporal can coexist with God's nature.

Time is a derivative property—a marker of deviation. Relativity describes clock behavior in a decaying system. It does not describe Infinum.

K.9 Dethroning Time to Reveal the Infinum

When time is treated as fundamental, the entire theological hierarchy is inverted:

- God becomes a participant in a larger system.

- Eternity is viewed as just a very long sequence.
- Creation is reduced to a clock.
- Sabbath is reduced to a tool.

The DPE Framework restores the correct hierarchy:

God's Nature → Eternity Infinum → Creation → Time (Only after Rupture)

Time is a local condition, not a cosmic ruler. Because Infinum is the default state:

- God is not inside time.
- God is not measured by time.
- God is not affected by decay.

Eternity as the default state is purer than giving time power over the Creator. Once time is understood as finite, derivative, and quarantine-specific, everything else becomes solvable: Why Eden had no decay, why the universe is terminal, and why the Sabbath is not tied to the spin of a planet.

K.10 The Identity of God and Eternity

If God and Eternity are one and the same, God cannot create anything non-eternal in His Original Architecture. This collapses centuries of confusion. Eternity is not a length or a duration. It is God's mode of being.

- God cannot enter time.
- God cannot use time.
- God cannot submit to time.

A Perfect, Eternal Being cannot produce finite, decaying things as His primary work. This is the part theologians avoid because it requires a forensic demolition of their current metaphors. If God is **Deliberate, Perfect, and Eternal**, then His creation must reflect those attributes.

- A perfect being cannot create imperfection.
- An eternal being cannot create finitude.
- A deliberate being cannot create accidental decay.

Time appears only when the system fractures. It is the evidence of rupture, not a tool of creation. Once time is dethroned, God's eternality becomes obvious. Once Infinum is recognized as the starting condition, the Original Architecture is revealed, and the purpose of the Sovereign Auditor to map the path out of the entropic hospice is fully realized.

ABOUT THE AUTHOR

Elvis Sanchez is an Axiomatic Independent, Structural Theorist, and the architect of the DPE (Deliberate, Perfect, Eternal) Framework. Operating outside the Satanic Architecture of Consensus, Sanchez specializes in the forensic audit of modern systems—identifying the structural contradictions within physics, mathematics, and the global Death Economy.

Refusing the Inherited Altitude of the academic lineage, Sanchez's work is grounded in Forensic Intuition and the removal of systemic deception. He is the author of a multi-volume forensic library that maps the regression of the modern world from the Original Architecture of the Creator into the entropic hospice of the Brilliant Fool.

He currently lives and works in New York, where he continues to monitor the propagation of the DPE Axiom through the Lineage Machine.

Elvis Sanchez rejects metaphorical speculation in favor of Logical Necessity. He operates as a Sovereign Auditor in New York City, providing a systematic recalibration for those who recognize the structural collapse of the modern consensus.

Inquiries: dpe.auditor@gmail.com

THE DPE FORENSIC LIBRARY

- *Deliberate, Perfect, Eternal*
- *Deliberate, Perfect, Eternal Part II: The Reckoning*
- *Infinum*
- *Infinum Spanish Edition*
- *Sin Mass: Dark Matter and the Architecture of Decay*
- *Math Idolatry: Canonization of Complexity*
- *Deception Priming*
- *Sin Mass: Spatial Medium and Regression*
- *Deathnology*
- *Unmask Satan: The Missing Operator*
- *Academic Atrophy: Falling Off the Shoulders of Giants*
- *Angel Physics*
- *Alien Inversion for Diversion*
- *Quarantine Cosmology: Einstein Meets Immutability*

THANK YOU

Thank you sincerely for reading Quarantine Cosmology: Where Einstein Meets Immutability. If the mentioned framework resonated with you, validated a deep intuition, or provided profound clarity, please consider leaving an honest review. Reviews remain essential for ensuring the forensic audit reaches the next observer who recognizes the structural collapse and seeks the DPE Axiom.

www.ingramcontent.com/pod-product-compliance
Lightning Source LLC
LaVergne TN
LVHW052211200926
848481LV00001B/73
* 9 7 9 8 2 3 2 8 2 6 6 8 0 *